The Green Future
Green Living and Taking Action on Climate Change

Leon

The Green Future: Green Living and Taking Action on Climate Change

Table of content

Chapter 1: Understanding Climate Change

1.1 Summary of Climate Change and Its Repercussions

One of the greatest problems facing humanity now is climate change. It has far-reaching effects on human cultures, economics, and ecosystems, in addition to the natural world. In this 1,000-word essay, we will study the causes and consequences of climate change, delving into the science underlying this complicated phenomenon and the impact it has on our world.

Climate Change Causes

The rise of greenhouse gases in the Earth’s atmosphere is the primary cause of climate change. By acting as a greenhouse, these gases contribute to global warming. There have been natural shifts in Earth's climate throughout millions of years, but the current trend of global warming is almost entirely attributable to human activity. The most significant contributors to global warming are:

First, CO_2 (carbon dioxide) emissions: The majority of manmade carbon dioxide emissions come from the combustion of fossil fuels like coal, oil, and natural gas for use in industrial processes and automobiles. The greenhouse effect is amplified as a result of the buildup of CO_2 in the atmosphere caused by these emissions.

2. Emissions of Methane (CH_4) When fossil fuels like coal, oil, and natural gas are extracted and transported, a significant amount of methane is emitted into the atmosphere. It is also emitted from cattle and other agricultural processes, as well as from the breakdown of organic waste in landfills.

(3) Deforestation: The elimination of forest cover, which serves as a carbon sink, adds to global warming. When trees are felled or

burned, the carbon they've stored in their tissues is released into the air.

Hydrofluorocarbons (HFCs), perfluorocarbons (PFCs), and sulfur hexafluoride (SF_6) are all released during certain industrial processes; these gases have a far larger warming potential than carbon dioxide (CO_2).

Agricultural Methods, Number Five Another powerful greenhouse gas, nitrous oxide (N_2O) is produced when synthetic fertilizers and intensive agriculture are used. Methane is also produced by the enteric fermentation of livestock.

6. Land Use Changes: Urbanization and other shifts in land use can have an impact on the Earth's surface albedo, which in turn affects how much sunlight is reflected or absorbed by the planet.
The effects of climate change are far-reaching, touching not just the natural world but also the international economy. Key effects include the following:

One, Heating Up Heatwaves are becoming more often and extreme as global temperatures continue to climb. The health of humans is put at risk, especially in disadvantaged communities.

2. Warming temperatures and melting ice: As a result of the melting polar ice caps and glaciers, sea levels are on the rise due to global warming. Because of this, millions of people could be forced to relocate when coastal areas get inundated and low-lying coastal villages are destroyed.

Thirdly, "Extreme Weather": Extreme weather events including hurricanes, droughts, floods, and wildfires are becoming more common and more destructive as a result of climate change. Communities and ecosystems are vulnerable to these kinds of disasters.

4. Ocean Acidification: Oceans around the world are absorbing excess CO2 and becoming more acidic as a result. This has severe repercussions for marine life, especially coral reefs and shellfish, as well as fisheries and coastal economies.

5. Dwindling Biodiversity Many plant and animal species are in jeopardy of extinction due to climate change and the resulting alteration of their habitats. Ecosystems can be thrown off balance, leading to a loss of biodiversity and the services they provide being compromised as a result.

6. Safety of Food and Water: Changes in temperature and precipitation patterns can affect crop production and water availability. This threatens the availability of food and water around the world and could lead to shortages and increasing competition for these essential commodities.

Effects on Health 7. Heat stress, the proliferation of infectious diseases, and the uprooting of entire populations are just a few ways in which climate change can have a negative impact on human health. The aged and the poor are among the most vulnerable groups in society.

Economic Implications The cost of climate change to the economy is high. Because of its widespread impact, prices may rise and economic stability may be threatened in industries as diverse as agriculture, insurance, and tourism.

Security on a National Scale: Resource scarcity and migration are two ways in which climate change can intensify or initiate conflict. This endangers the safety of the entire world.

The Social and Political Consequences Particularly in already-vulnerable countries, climate change has the potential to generate or exacerbate preexisting social disparities and cause political unrest.

The repercussions of climate change are far-reaching and penetrate every facet of our lives, and they are all interconnected. Reducing greenhouse gas emissions, adjusting to the changes that are already occurring, and increasing international collaboration to mitigate the worst impacts are all necessary to address these repercussions.

Conclusion

Human-caused climate change is a serious issue on a global scale. Its origins in the emission of greenhouse gases are well-known, and its effects on the world are becoming increasingly apparent. To protect the future of Earth and its inhabitants, addressing climate change is not only an environmental necessity, but also a moral and ethical obligation.

Mitigating the causes of climate change by lowering greenhouse gas emissions is vital. Reforestation, sustainable agriculture methods, and waste reduction are all positive developments that can help the planet. Moreover, protecting vulnerable populations and ecosystems requires adjusting to the changes that are currently happening.

The first step in tackling this global crisis is realizing the effects of climate change and the factors that contribute to it. To preserve Earth for future generations will take concerted effort on a global scale, a shift in policy, and personal initiative. The stakes of inaction are too high to be ignored at this time.

1.2 The Critical Need to Act on Climate Change Around the World.

Climate change is a formidable obstacle that has never before been faced by humanity. We must act quickly to solve this worldwide problem. This dilemma goes well beyond national boundaries and threatens the stability of our planet's ecosystems and human communities. The worldwide urgency of tackling climate change, the repercussions of inaction, and the necessity of quick and concentrated actions to counteract it will all be discussed in this 1,000-word essay.

The effects of climate change are already being felt now. Glacier melting, higher sea levels, more frequent and severe weather events, and changing climate patterns are all telltale signals. Numerous reports by the Intergovernmental Panel on Climate Change (IPCC) give scientific evidence and consensus on the fact, causes, and implications of climate change. These reports serve as a rallying cry, warning that time is of the essence as the crisis worsens by the degree.

What Will Happen If We Do Nothing

The results of doing nothing about climate change are severe, and in some cases, irrevocable. As a result, they pose threats to the world's ecosystems, economies, populations, and future generations. Let's examine many of the most important results:

Impacts on the Environment 1. Ecosystems are collapsing because of climate change. Deforestation and the loss of habitats provide a hazard to flora and fauna on land, while ocean acidification poses a threat to marine life. Ecosystems as a whole may be thrown off by the loss of species variety.

The Rise of Sea Levels: Warming temperatures cause ice caps and glaciers to melt, which in turn causes sea levels to rise. Saltwater intrusion, erosion, and elevated flood risks are already having an impact on coastal towns.

Third, Extreme Weather Events, such as Hurricanes, Droughts, Floods, and Wildfires, are becoming more often and more intense, wreaking havoc on people's lives and economies and causing extensive damage and loss.

The availability of both food and water is impacted by climate change. Reduced agricultural yields due to variations in precipitation and temperature can cause food shortages and boost competition for scarce resources.

Impacts on Health 5. Heat-related ailments, disease transmission, and the worsening of existing medical disorders are all possible outcomes of a warming planet. Most at risk are those of society who are already weak.

Economic Implications: The cost of climate change to the economy is high. Insurance companies and farmers alike may feel the financial impact of infrastructure and property damage and agricultural losses.

Security at Home and Abroad, Number Seven Climate change can intensify conflicts by contributing to resource scarcity and migration. As climate-related conflicts can have far-reaching effects, they represent a danger to both domestic and international stability.

Social Inequalities are exacerbated by climate change. It disproportionately affects already-disadvantaged groups and widens existing gaps in society.

The Urgent Need to Take Measures

It is impossible to overstate the importance of taking action to stop climate change. The longer we wait to take action, the harder and more expensive it will be to lessen the damage. There are a number of pressing reasons why we must act now:

1. Falling Off: There may be thresholds in the Earth's climate system beyond which changes become irreversible. A catastrophic rise in sea level would occur, for instance, if the Greenland Ice Sheet melted dramatically.

The Financial Impact Costs associated with climate change adaptation and mitigation may rise if action is delayed. The cost of inaction is much higher than the price tag on a sustainable, low-carbon future.

International Partnership The effects of climate change are felt on every continent. To adequately address this issue, international cooperation is required. Lack of prompt action reduces confidence and complicates efforts to work together.

Responsibility in Moral and Ethical Matters, We have an ethical and moral obligation to preserve Earth for future generations in our role as the planet's stewards. It would be a betrayal of this duty not to act.

5. Creative Answers The adoption of green technologies and sustainable practices can contribute to economic growth and the creation of new jobs if immediate action is taken.

Individual Responsibility

Every person on Earth can make a difference in the fight against climate change. While systemic shifts are essential, the sum of individual efforts can have a much greater impact. Individuals can help in a variety of ways, including:

One way to lessen one's carbon footprint is to adopt energy-saving habits, use eco-friendly modes of transportation, and curb wasteful spending.

Second, "Advocate" and "Educate": Engaging in climate education and communication activities; advocating for climate-friendly policies; increasing public understanding of climate change.

Choosing renewable sources of energy, recycling more, and cutting down on food waste are all examples of sustainable lifestyle choices.

Taking part in community-level climate initiatives, clean energy projects, and campaigns for ecologically responsible policies is an example of what we mean by "Community Engagement" (also known as "4").

5. Consumer Influence: Buying from sustainable businesses and speaking up for businesses to do the right thing for the environment.

Policy and Governmental Response

The global urgency of climate change demands collective action. Developing a holistic strategy requires the support of government policies and international agreements.

The shift to clean energy sources should be a top government priority, and governments should offer financial incentives for the development of clean energy technology.

Second, Carbon Pricing: putting in place carbon pricing mechanisms to show the real cost of carbon emissions and motivate cuts.

Thirdly, Regulations and Standards: Creating and implementing environmentally friendly rules and regulations is essential to fostering long-term viability and cutting down on pollution.

Lastly, we can all work together to combat climate change by supporting and taking part in international agreements like the Paris Accord.

5. Investment in Research and Innovation: investing in green technology R&D to speed up the shift to a low-carbon economy that is more sustainable.

Conclusion

There is no greater problem facing humanity now than the urgent need to combat climate change on a global scale. Neglecting this issue will have severe repercussions for our planet, our economies, our health, and our communities. There must be immediate response on the part of individuals, groups, states, and nations. We can build a sustainable and resilient future despite climate change, but it will take a committed and persistent commitment from all of mankind. We owe it to ourselves and to the people who will live on this earth when we're gone.

Chapter 2: Personal Carbon Footprint

2.1 A Roadmap to Climate Responsibility Through Self-Assessment of Your Carbon Footprint

It has never been more important to learn about and evaluate your personal carbon footprint than in today's day, when climate change is one of the most severe global concerns. The quantity of carbon dioxide (CO2) and other greenhouse gases released into the atmosphere as a result of human activities can be quantified by looking at what is known as a carbon footprint. Everyday things we do, like getting in a car or turning on the lights, add up to a bigger lifestyle that includes the foods we consume and the things we buy. In this 1,000-word essay, we'll talk about why it's so important to figure out your personal carbon footprint, how to do it, and what you can do to lessen your personal environmental impact.

The Importance of Humans' Carbon Footprints

Your carbon footprint is an ecological signature that reflects the effects of your lifestyle on the environment. It's a must-have for any serious attempt to comprehend and counteract global warming. The importance of calculating and comprehending one's carbon footprint is highlighted by the following points:

1. Awareness: Measuring your carbon footprint improves awareness about the environmental effects of your actions. You may learn how much of an impact you have on global warming by using this tool.

Responsibility 2. Taking personal responsibility for your emissions is made easier when you are aware of your carbon footprint. With this information in hand, people may make greener decisions.

Goals for minimizing your carbon footprint can be established when a baseline has been established. It lays up an unmistakable foundation for eco-friendly lifestyles.

4. Climate Action: You can do your part in the fight against global warming by calculating your personal carbon footprint. Individual efforts, when added to those of the entire population, can have a major impact.

Environmentalism in the Classroom: You can learn a lot about the effects of your daily actions on the environment by estimating your carbon footprint.

Your carbon footprint is a numerical representation of your contribution to global warming, measured in terms of carbon dioxide equivalent (CO2e). It includes things like how much energy you use, how you get around, what you eat, and how much stuff you buy. The steps to determining your personal carbon footprint are as follows:

Energy Usage is the first place to look. Find out how much carbon dioxide you're adding to the atmosphere from your home's use of energy and gas. The vast majority of electric companies supply either this information or methods to approximate it.

2. transit: Think about how you'll get to and from work each day, if you'll be flying, and if you'll be taking public transit. The amount of greenhouse gases produced by various transportation options varies.

Thirdly, consider the ecological effects of your diet. The environmental impact of a plant-based diet is typically smaller than that of a diet heavy in animal products.

4. Waste Generation: Determine how much garbage you produce and how much of it is recycled vs thrown away. Your carbon footprint increases due to landfill methane emissions.

5. Consumer Goods: Think about what you buy, how it's made, and how it gets to you. Emissions are produced at every stage of a product's life cycle, from mining to recycling.

Include any extra emission-causing actions, such as flying, recreational activities, and hobbies under the heading "Other Activities."

When you have collected this information, you may utilize carbon calculators online or talk to professionals to get an idea of your overall carbon footprint. Keep in mind that different carbon footprint calculators will produce different answers based on the quality of the information you provide.

Lessening Your Impact On The Environment

The next step after calculating your carbon footprint should be to implement real changes to lessen it. Here are a few approaches to think about:

1. Energy Efficiency: Make your home more energy efficient by installing solar panels or other renewable energy sources, improving the insulation, and adopting energy-efficient equipment.

Secondly, Transportation Share rides, take the bus, ride your bike, or walk to work to lessen your impact on the environment. Consider purchasing an electric or hybrid vehicle.

Reducing meat intake, increasing plant-based foods, and favoring local and seasonal produce are all ways to adopt a diet with a smaller carbon footprint.

4. Waste Reduction: Reduce your trash by reusing, recycling, and composting as much as possible. Refrain from using anything more than once.

Fifthly, Consumer Options: Choose items that won't harm the environment when you shop for them. Help businesses that care about the environment by purchasing their products and using their services.

6. Flights Avoid flying whenever possible in favor of teleconferencing or staying put. If you absolutely must take a flight, look into carbon offset options.

7. Advocacy: Promote climate-friendly policies and back regional, state, and federal initiatives to curb global warming.

8. Teach Others: Discuss the issues of carbon footprints and global warming with those closest to you. Inspire them to follow in your footsteps.

Issues and Things to Think About

There are several factors to take into account while calculating and attempting to lessen one's carbon footprint. Some things to keep in mind are as follows:

First, a Balance Act Sometimes, you have to choose between comfort and sustainability if you want to reduce your carbon footprint. Finding the correct balance is key.

The Economic Climate: In the short term, certain environmentally beneficial choices may cost more. However, they can have positive financial and ecological consequences in the long run.

Thirdly, Availability of Various Options: Where you live can have a big impact on your access to eco-friendly options like public transit and organic groceries. Your efforts to lessen your impact on the environment may be hampered.

4. Behavior Change: It might be difficult to make the changes in behavior and habits that are necessary for sustainable life.

5. A World View: Changes at the political and corporate levels are just as crucial as changes at the individual level. It's essential to push for systemic reform.

Carbon Offsets, Number Six: To offset their carbon footprint, some people choose to finance "carbon offset" projects. Environmental initiatives like tree planting and the creation of alternative energy sources are supported by these schemes.

Conclusion

Taking the time to calculate your personal carbon footprint is a major move toward environmental responsibility. You'll be able to lessen your negative effects on the planet thanks to the knowledge you gain. Although people's efforts are needed, they shouldn't be the only ones. To combat this worldwide issue, people, organizations, governments, and corporations must work together. Learning about and cutting down on your personal carbon footprint is an important step toward a more sustainable and accountable future as we fight to lessen the effects of climate change. This is an endeavor that has far-reaching implications for the health of our planet and the lives of generations yet unborn.

2.2 The Responsibility of Individuals in Countering Climate Change

The effects of climate change on the natural world, the economy, and human society are far-reaching, making it one of the most pressing problems of our day. The individual's contribution is crucial, even though solving climate change requires global cooperation and structural improvements. In this 1,000-word article, I want to discuss the necessity of personal action and collective responsibility in bringing about a more sustainable future in the face of climate change.

Recognizing the Climate Emergency

Individual responsibility for climate change mitigation should follow a thorough examination of the climate crisis. Greenhouse gases in the Earth's atmosphere are the primary cause of climate change. These gases, especially carbon dioxide (CO_2), cause the Earth to warm because they act as a blanket for absorbed solar radiation. Climate is affected by both natural and human forces, but the recent acceleration in warming is mostly attributable to the latter. Greenhouse gases are primarily caused by:

Fossil fuels (1) Most carbon dioxide (CO_2) emissions come from the combustion of fossil fuels (such as coal, oil, and natural gas) for use in industrial processes and automobiles.

Methane (CH_4) and nitrous oxide (N_2O) are produced via agricultural techniques such as enteric fermentation in livestock and the use of synthetic fertilizers.

3. Deforestation: The cutting down or burning of trees releases carbon that had been stored in the trees and served as a carbon sink by absorbing CO_2.

Hydrofluorocarbons (HFCs), perfluorocarbons (PFCs), and sulfur hexafluoride (SF6) are all gases with a high global warming potential that are released during specific industrial processes.

The decomposition of organic waste in landfills is a major source of methane, a potent greenhouse gas.

The first step toward combating climate change is becoming aware of these emission causes.

The Impact of One's Own Efforts

The magnitude of the climate catastrophe can make it difficult to see how one person can make a difference, but the truth is that everyone does. Here are a few things you can do right now to help slow global warming:

The use of LED light bulbs, insulation upgrades, and programmable thermostats are all examples of energy-efficient home improvements that can help cut down on utility bills and carbon emissions.

To lessen the environmental impact of getting to work, many people are turning to sustainable modes of transportation like public transportation, carpooling, bicycling, walking, and electric cars.

Emissions from livestock and farming can be greatly reduced if people adopted plant-based diets, ate less meat, and favored seasonal, locally-grown produce.

4. garbage Reduction: Minimizing garbage, recycling, composting, and minimizing single-use plastics helps cut emissions from landfills.

Ethical shopping for the environment Choosing products with environmental impact in mind can encourage greener production methods and lessen your personal carbon footprint.

Advocating for climate-friendly legislation and educating yourself and others about the issue are great ways to help bring about systemic change.

7. Water Conservation: Reducing water use at home and pushing for water-saving methods can decrease the energy required for water heating and distribution.

Clean energy is promoted through individual and community activities that support and invest in renewable energy sources like solar and wind power.

The Power of the People and Local Initiatives

Although effective on their own, individual efforts are more potent when they are part of a larger group effort. Movements from the bottom up, local initiatives, and international campaigning are all crucial to bringing about structural shifts.

1. Eco-Group Membership Sharing ideas, working together on projects, and advocating for change are all made possible through membership in local or online environmental groups.

Local Clean-Up Efforts: Participating in clean-up efforts in your neighborhood is a great way to help the environment and develop a feeling of civic duty at the same time.

The third method to affect systemic change is to vote for political leaders and policies that put climate action first.

4. Lobbying and Political Engagement: Meeting with, writing to, and calling legislators at both the state and federal levels can help promote climate-friendly laws.

Supporting Eco-Friendly Policies 5. Change can be sparked at the neighborhood level by advocating for green infrastructure, renewable energy initiatives, and sustainable habits.

Solidarity on a Global Scale: Individuals can promote international collaboration and assist efforts to address the global scope of climate change by raising awareness of the issue and advocating for action.

Innovations and Green Technologies

Green technology and innovation developments play an important part in climate change mitigation alongside individual and community actions. These developments can be found in fields as diverse as energy, transportation, agriculture, and construction. Such instances include:

First, renewable energy sources such as solar and wind, combined with improvements in energy storage technologies, are decreasing the need for fossil fuels in the creation of electricity.

Electric Vehicles, Second: Emissions from transportation have decreased thanks to the growing popularity of electric vehicles and improvements in charging infrastructure.

Energy-efficient designs, green roofs, and passive heating and cooling systems are just a few examples of sustainable building practices that reduce a building's carbon footprint during construction and use.

4. Agricultural Innovations: Emissions from agriculture can be reduced by the adoption of sustainable farming practices, precision agriculture, cover crops, and crop rotation.

5. Green Startups: Sustainability-oriented enterprises, such as circular economy companies and renewable energy startups, are at the forefront of green industry innovation.

The Importance of Raising People's Awareness

The dissemination of knowledge is essential in the fight against global warming. The first step is to educate oneself about the science of climate change, its effects, and possible remedies. Additionally, raising awareness in your community, through social media, forums, and grassroots efforts, can help mobilize people to take action. By raising people's level of knowledge and consciousness, we can all make better, more well-informed decisions.

Issues and Things to Think About

Despite the unquestionable importance of individual action in combating climate change, there are obstacles and factors to bear in mind.

One of the most difficult aspects of adopting a sustainable lifestyle is overcoming ingrained patterns of behavior.

While going the eco-friendly route may have a larger initial investment, it usually pays off in the long run through cost savings and less environmental impact.

Thirdly, Access to Alternatives: Your capacity to lessen your environmental impact is influenced by the ease with which you can access sustainable options like public transportation and organic meals.

4. Group Effort Changes at the political and corporate levels are just as essential as changes at the individual level. It's essential to push for systemic reform.

5. A World View: Climate change is a problem that affects people all around the world. To properly solve the problem, international cooperation is required.

Conclusion

The individual's part in reducing greenhouse gas emissions is crucial to international efforts to address this urgent problem. The cumulative effect of individuals' efforts on their neighborhoods, regions, and countries is tremendous. While combating climate change involves structural changes and societal responsibilities, individual acts are crucial. Every day, through your actions, choices, and advocacy, you help build a more just and sustainable future. The problem of climate change is one that requires a

To create a more sustainable environment for future generations, we need a collective reaction, and each person must play a part.

Chapter 3: Green Living at Home

3.1 Energy-efficient home improvements

To combat climate change, our houses must be more than just a place to sleep. Making your home more energy efficient is a great way to lessen your impact on the environment, save money, and feel more at ease in your own home. In this 1,000-word paper, we will discuss the need of making sustainable changes to one's home, the variety of options for doing so, and the financial and ecological benefits of doing so.

Acquiring an Awareness of Energy Efficiency's Significance

To be energy efficient means to reduce one's home's energy use without compromising on either performance or comfort. There are many compelling arguments in favor of boosting energy efficiency:

First, "Climate Change Mitigation": Greenhouse gas emissions, a major contributor to global warming and climate change, can be greatly reduced by cutting energy use in the household.

2. Decreasing Utility Bills: Because energy-efficient homes use less electricity and gas, their owners enjoy cheaper utility bills.

3. Improved Comfort Newer, more energy-efficient homes typically have better insulation, ventilation, and climate control.

4. Increasing Property Value: Homes that are both comfortable and environmentally friendly are in high demand. Property value can be increased by investments in energy efficiency.

Energy security and price stability are aided by reducing reliance on fossil fuels, which is another benefit of reducing energy consumption.

Options for Making Your Home More Energy Efficient

To make a home more energy efficient, homeowners can choose from a wide variety of options depending on their preferences and budgets. Among the most typical alterations are:

Having well-insulated walls, ceilings, and floors is crucial for maintaining a comfortable temperature within a building. High-quality insulation can lessen the workload of your HVAC system.

Energy-Efficient Windows and Doors Heat loss and drafts can be avoided by putting in double-glazed windows and properly sealing the doors. This improves convenience and reduces utility bills for both heating and cooling.

Keeping a building at a suitable temperature requires a lot of energy, but modern HVAC systems are made to be more energy efficient.

Fourth, LED Lighting can minimize both electricity use and the frequency of bulb replacements by replacing incandescent and fluorescent lamps.

Smart thermostats, number 5. Smart thermostats help homeowners save money on their utility bills by allowing them to better regulate the temperature in their homes while they are away.

Solar Panels, number 6. In order to generate power, solar photovoltaic (PV) panels use sunlight instead of fossil fuels.

Seven, Energy-Efficient Home Appliances: Appliances like refrigerators, washing machines, and dishwashers can be upgraded to use much less electricity if they are Energy Star certified.

Insulation is improved, drafts are eliminated, and energy is saved by locating and sealing air leaks in the building exterior.

Heat recovery ventilation (HRV) systems are 9th on the list because they improve energy efficiency and indoor air quality simultaneously.

It is possible to reduce the cost of water heating by upgrading to an energy-efficient water heater, insulating the hot water pipes, and adjusting the water heater's temperature to a lower setting.

Green and cool roofs, number 11. The use of air conditioning can be reduced and internal temperatures lowered by installing green or cool roofs on a structure.

The Twelve Advantages of Renewable Energy: In addition to solar panels, other sustainable energy sources that homeowners might look into include wind turbines and geothermal heating and cooling systems.

Positive Effects on the Economy and the Environment

There are numerous financial and environmental gains that can be realized through making homes more energy efficient.

First, Money Saved: Perhaps the most immediate advantage is the reduction in utility expenditures. Homes that are more energy efficient use less power, which means homeowners save money each month.

Second, upgrading to more energy-efficient features can increase a home's resale value and make it more appealing to purchasers. Homes that are more energy-efficient typically sell more quickly and for a higher price.

Thirdly, many governments provide financial incentives and tax credits to encourage energy-efficient house improvements, substantially reducing the initial costs of changes.

4. Reduced Maintenance: Energy-efficient systems and materials frequently require less maintenance and have longer lifespans, saving homeowners money on repairs and replacements.

Better insulation, more accurate temperature regulation, and fewer drafts all contribute to a more pleasant indoor climate.

6. Impact on the Environment: Reducing energy use aids in the fight against climate change by decreasing emissions of greenhouse gases. Homes that use less energy to run are better for the planet.

7. Power Availability: By decreasing energy consumption, we can lessen our reliance on fossil fuels and help keep energy costs down.

Increasing ventilation and filtering pollutants are two ways that energy-efficient upgrades like HRV systems improve indoor air quality.

9. Less Stress on the Grid: Energy-efficient households put less burden on the electrical grid since they use less energy overall. This has the added benefit of reducing the frequency and severity of power outages.

The Obstacles and Opportunities on the Road to a Sustainable Lifestyle

There are a lot of reasons to make your home more energy efficient, but there are also some things to think about.

Initial Expenses: The upfront costs of certain energy-saving measures may be prohibitive. Long-term savings and government incentives must be factored into any assessment of these investments, though.

Two, ROI Calculation: Homeowners should evaluate the return on investment (ROI) for energy-efficient improvements, taking into account energy savings and potential property value gains.

3. Different Climate Requirements Depending on the weather, different areas may need different energy-saving improvements. In colder climates, insulation may be a top priority, while in warmer regions, cooling efficiency may be the priority.

Upgrade Bundling 4: Combining several different energy-saving measures usually results in the best overall savings. This may call for some advanced preparation and collaboration.

Lifestyle adjustments, such as becoming more cognizant of energy use or adopting energy-efficient habits, may be necessary to achieve energy efficiency.

Professional Guidance: Homeowners can find the best upgrades for their homes by consulting with energy auditors or home performance consultants.

Energy efficiency is the future and the way forward.

Upgrades to make homes more energy efficient are more than a passing fad; they signal a cultural change toward a more eco-friendly way of life. With increasing awareness of the climate catastrophe and the need to cut carbon emissions, these improvements are urgently needed. To ensure the future of our planet, we must all do our part by making the switch to more energy-efficient homes.

More and more homeowners are realizing the benefits of energy efficiency, driving up the demand for cutting-edge innovations and eco-friendly methods. Governments, companies, and communities are all part of this transforming journey. Investing in renewable energy and other forms of energy efficiency is a great way to reduce your carbon footprint and save money in the long run. It's a way forward that can improve people's lives now while also making the world a better place for future generations to live in.

3.2 The Road to Environmental Stewardship Is Paved With Water Conservation and Efficiency.

All life on Earth depends on water, making it a precious resource. Adopting sustainable water usage and waste reduction measures has become vital as the world faces growing worries over water scarcity and environmental deterioration. In this 1,000 word paper, we will discuss the importance of water sustainability, the consequences of water waste, and methods for protecting and conserving this precious resource.

Realizing the Value of Water Conservation

It's easy to take water for granted when it's plentiful, as is the case in many parts of the world. Unfortunately, water is a limited resource, and access to it varies widely. There are many reasons why it's critical to have a firm grasp on the concept of water sustainability:

1. Environmental Conservation: The loss of freshwater habitats and the deterioration of water bodies are direct results of the overuse and contamination of these resources.

Alterations to the Climate: Climate change exacerbates water scarcity and fluctuating precipitation patterns. Sustainable water practices help mitigate and adapt to these changes.

Thirdly, Energy Efficiency: Water and energy are inextricably intertwined. Water conservation is a key factor in reducing the amount of energy needed for water treatment and distribution.

4. Economic Savings: Lower water costs are the direct result of less water being wasted at both homes and businesses. Water treatment and infrastructure costs are also reduced.

5. Health and Sanitation: Sustainable water consumption practices promote public health by minimizing the spread of waterborne diseases and guaranteeing universal access to clean drinking water and well-maintained restrooms.

Global Water Security, Number Six As the global population continues to rise, competition for freshwater supplies intensifies. The safety of the world's water supply depends on the adoption of sustainable water practices.

The Consequences of Water Wastage

Wasted water has far-reaching effects on ecosystems, communities, and economy. The following examples highlight the effects of water waste:

1. Resource Depletion: Friable freshwater resources are finite and, because to wasteful exploitation, have become scarce in some locations.

Second, there is the issue of Energy Consumption; both the processing and delivery of water require a lot of power. Energy needed for water treatment and transportation is also lost when water is wasted.

Thirdly, there are Economic expenses associated with water waste, including higher water bills for customers, greater government infrastructure expenses, and significant agricultural economic losses.

4. Damage to the Environment: The existence of many freshwater species is in jeopardy because of the overexploitation of water sources like wetlands.

5. Infrastructure Strain: Unsustainable water consumption places a burden on the system that delivers water, raising the probability of damage to pipes and water loss.

6. Drought Vulnerability: In drought-prone areas, water waste can increase water shortages and reduce agricultural output.

Strategies for Sustainable Water Usage

Individuals, communities, corporations, and governments all have a part in the shift to more sustainable water usage. Several methods are outlined below for establishing water conservation measures that will last:

1. Domestic Water Conservation

- Leak Repairs Make it a habit to inspect your plumbing, sinks, and showers for leaks on a regular basis.

Low-flow toilets, high-efficiency washing machines, and water-saving dishwashers are just a few examples of water-saving appliances that should be installed.

- Shorter Showers: Take shorter showers and install a water-saving showerhead.

Rainwater can be collected and stored in rain barrels for later use in the garden or other outdoor applications.

Landscape sustainably by using drought-tolerant plants and other xeriscaping techniques.

To save the most water, only use full loads in your washing machine and dishwasher.

2. Farming Methods and Water Management:

Reduce water waste in farming by using effective irrigation techniques like drip watering.

Select crop varieties that work well in the soil and with the available water.

- Soil Quality: Investing in healthy soil will increase water retention and decrease irrigation needs.

Thirdly, Business and Industrial Water Management

- Water Audits: Conduct frequent water audits to detect and address inefficiencies in industrial and commercial water use.

Water recycling and reuse solutions should be integrated into all industrial processes.

Designing and constructing buildings with low-water-use appliances and systems counts as one of the green construction practices.

4. Collecting Water from the Rain:

Rainwater can be collected and stored for use in non-drinking applications such as cooling systems, landscape irrigation, and toilet flushing.

5. Informing and Educating the General Public:

Individuals and communities can be educated and inspired by public campaigns and educational initiatives that promote water conservation.

6. Laws and Regulations Enforced by the Government:

Governments can encourage the use of water-saving devices and practices by enacting water-conservation regulations, establishing efficiency requirements, and offering financial incentives.

Water Reuse and Reclamation 7.

Reduce the negative effects of wastewater discharges on the environment by investing in modern wastewater treatment and recycling technology.

Desalination, number 8.

- In places with limited freshwater resources, desalination technology can provide an extra source of clean water.

Sustainability and waste minimization methods

Sustainable water use is intrinsically tied to waste reduction. Conservation of resources and preservation of the natural world are aided by minimizing water waste in a variety of contexts. Important methods for cutting down on trash include:

Reduce the amount of water and nutrients lost due to over-irrigation in farming and landscaping.

runoff. Monitoring soil moisture and employing more precise irrigation techniques can help cut down on wasteful water use.

Water from sources like the kitchen sink, the bathroom shower, and the laundry machine (together referred to as "greywater") can be recycled for non-drinking uses like watering the garden or flushing the toilet.

Third, for the sake of preventing stormwater pollution and making better use of captured rainfall for irrigation, you should put into place sustainable stormwater management methods.

Industry should emphasize water-efficient operations and wastewater treatment to reduce water waste and pollution, per [Responsible Manufacturing] principle 4.

Fixtures and appliances that use less water Water waste can be greatly reduced with the installation of water-efficient fixtures and appliances in homes, businesses, and factories.

6. Wastewater Recycling: Effluents are discharged into natural water bodies less frequently because of advanced treatment

technologies that recycle and reuse treated wastewater for various applications.

Issues and Things to Think About

Despite the undeniable upsides of waste minimization and water conservation, there are still several obstacles to bear in mind:

First, it can be difficult to alter one's behavior and establish new routines in order to adopt more water-friendly activities.

Investment Expenses (2): Some sustainable water technologies and practices may involve upfront investments, but they can result in long-term savings.

Thirdly, Infrastructure Upgrades: It can be both time-consuming and expensive to retrofit older buildings and infrastructure to be more water-efficient.

4. International Dispersion Climate, water availability, and infrastructure differences can have a substantial impact on the water problems and potential solutions in every given area.

5. Support in Lawmaking and Regulation: Through laws, incentives, and regulations, governments play a crucial role in fostering sustainable water practices. Advocating for such assistance is vital.

Conclusion

Responsible environmental stewardship and a more sustainable future depend on efficient water use and minimal waste production. It is critical for individuals, communities, organizations, and governments to embrace water-saving practices and waste-reduction initiatives as the world struggles with increasing water scarcity and environmental concerns. In addition to preserving a limited resource, these measures also benefit the planet, the economy, the ecology, and the fight against climate change.

For the sake of our world and the generations to come, we must all do our part to make the switch to more sustainable water practices. It will take a concerted effort on all levels, from the individual to the international, to solve the world's water problems and keep water useful as a resource for life and a better, more sustainable planet.

Chapter 4: Sustainable Transportation

4.1 Sustainable Transportation Alternatives: Charting the Course for Tomorrow

The ways in which we commute have significant effects on our ecosystem, economy, and quality of life. It has never been more important to carefully consider your commute options than it is now, as the world struggles with climate change and environmental destruction. Cleaner air, less traffic, and a smaller carbon impact are all possible with eco-friendly commute options. In this 1,000-word paper, we will discuss the importance of sustainable transportation practices, the range of accessible solutions, and the game-changing effects of going green on the daily commute.

Recognizing the Value of Eco-Friendly Transportation

Reasons why eco-friendly transportation options are essential include:

1. Environmental Stewardship: The use of fossil fuels in conventional modes of transportation is a major contributor to greenhouse gas emissions and air pollution, both of which have severe negative consequences for the environment and for people's health.

The transportation sector is a major contributor to global warming. We can all do our part to fight global warming by choosing more eco-friendly modes of transportation.

Thirdly, Improvement to Air Quality occurs because people who commute in a sustainable manner produce less of the hazardous pollutants that worsen air quality, hence decreasing the prevalence of respiratory ailments and other health problems.

Economic Benefits 4. Lower fuel prices, less traffic congestion, and enhanced public transportation are just a few ways that opting for environmentally friendly modes of transportation may save money for individuals and communities.

5. Decreasing Traffic Congestion Reduced commute times, less stress, and increased road safety are all benefits of encouraging eco-friendly modes of transportation.

Alternative, Environmentally Friendly Methods of Commute

The term "eco-friendly commuting" refers to a wide range of travel options with the common goal of minimizing negative effects on the environment. There is a broad spectrum of accessibility, availability, and adaptability among these choices. Common environmentally friendly modes of transportation include:

Cycling Bicycles are a healthy and eco-friendly way to go about town. The environmental benefits of cycling are complemented by the health benefits of cycling. To make cycling a more viable transportation choice, several urban centers are constructing bike lanes and other amenities.

(2) When Walking: Walking is the easiest, most sustainable, and most accessible form of transportation. This emission-free alternative has many positive effects on human health.

Thirdly, using public transportation like buses, trams, subways, and trains can help you reduce your carbon footprint while moving a huge number of people. Sustainable development demands continued funding for public transportation systems.

Fourthly, Ridesharing and Carpooling: By reducing the number of cars on the road, carpooling helps reduce pollution and saves money on gas.

The Electric Vehicle (EV) Market: Cleaner than vehicles powered by internal combustion engines include electric automobiles, bikes, and scooters. As the technology behind EVs improves, people will have easier time adopting this mode of transportation.

Reduced fuel use and pollutants are a benefit of hybrid vehicles, which use a combination of an internal combustion engine and an electric motor.

7. Car-Sharing Services: Car-Sharing platforms provide access to automobiles just when needed, lowering the requirement for individual car ownership and increasing the effectiveness of available resources.

Telecommuting, number eight: Telecommuting or working from home eliminates the need for daily commuting, a major contributor to both traffic and pollution.

Scooters and bicycles with electric motors: Short-distance electric scooters and bicycles are a practical and sustainable substitute for autos.

10. Trains and Light Rail: High-speed trains and light rail systems are efficient and sustainable for long-distance and urban transportation, respectively.

11. Green Fuels: Compared to conventional gasoline and diesel, vehicles that run on biofuels, hydrogen, or other sustainable fuels produce fewer emissions.

Solution(s) for Intelligent Mobility: Smart mobility solutions have emerged as a result of developments in technology and urban design, and they include things like ride-hailing applications, real-time traffic information, and navigation tools that optimize routes to minimize emissions.

The Game-Changing Effect of Eco-Friendly Commute Options

Positive changes in environmental, economic, and social contexts can result from adopting sustainable commute practices.

1. Benefits to the Environment Sustainable commuting helps make the world a better place by cutting down on harmful emissions and pollution. It's a climate change solution that safeguards the planet's natural ecosystems and species at the same time.

Economic Savings 2. Eco-friendly commuting can result in substantial personal and community economic savings. If people drove less and used public transportation more, they could save money on gas, maintenance, and car payments.

Thirdly, [[Health and]] [[Welfare]:] Physical fitness, decreased stress, and improved mental health are all benefits of eco-friendly commute options like bicycling and walking. The health benefits of lower emissions and thus improved air quality are obvious.

The time saved by using sustainable methods of transportation instead of driving can be better spent with loved ones or on hobbies.

5. Improved Urban Planning: A move toward sustainable commuting pushes urban planners to invest in infrastructure that prioritizes pedestrian- and bike-friendly urban design, public transportation networks, and better traffic management.

6. Community Building: Sustainable commuting creates a sense of community by encouraging the use of shared spaces and facilitating interaction between neighbors.

7. Equity and Accessibility: Sustainable commuting solutions give equal access to transportation for all individuals, including those who may not have access to a personal vehicle.

8. Less Reliance on Fossil Fuels: Changing how we get around means less reliance on fossil fuels, which are both limited and polluting. This improves energy security.

Issues and Things to Think About

The advantages of eco-friendly transportation are clear, but there are also some things to bear in mind.

1. Investment in Infrastructure Cycling lanes, public transportation systems, and electric vehicle charging networks are just a few examples of the kinds of infrastructure that are needed to support more environmentally friendly modes of transportation.

2. Behavior modify: It might be difficult to get people to modify their behavior and habits when trying to get them to embrace sustainable commuting techniques.

Thirdly, Disparities in Transportation: There may be inequalities in the development and availability of infrastructure, meaning that not all regions have access to the same sustainable commuting alternatives.

4. Economic Barriers: Some people may be put off by the higher initial expenses of eco-friendly commuting solutions like electric vehicles.

5. Support in Lawmaking and Regulation: Governments have a significant role in encouraging and supporting eco-friendly commuting through laws, incentives, and regulations. Advocating for such assistance is vital.

Conclusion

The fight against climate change, the reduction of pollution, and the building of more sustainable and fair societies relies heavily on eco-friendly transportation. The environment, health, and quality of life are all affected by the decisions we make on a daily basis throughout our commutes.

Sustainable transportation choices can't just be advanced through individual efforts; governments, businesses, and communities must all play a substantial part as well. To commute in a way that is good for the environment and future generations requires more than simply a personal decision. The way we travel through the world is not just a question of convenience as we work towards a more sustainable future; it is also a reflection of our commitment to environmental stewardship and creating a world that is healthier, more equal, and more sustainable for everyone.

4.2 Greener Ways to Travel: Strategies for the Future

Travel is an essential component of the human experience, providing possibilities for discovery, adventure, and cultural interaction. However, as the world's population rises and environmental problems multiply, we must seriously evaluate how our travel choices affect the earth. Responsible tourism is a win for both local economies and the environment, and sustainable travel practices are one way to make that happen. In this 1,000 word essay, I want to delve into the relevance of sustainable travel, investigate several sustainable vacation possibilities, and talk about the revolutionary potential of embracing eco-conscious tourism practices.

Acquiring an Awareness of the Value of Eco-Friendly Vacations

The goal of "sustainable travel" is to see the globe while leaving as little of a footprint as possible on the places we visit and the people we meet. The following details highlight its significance:

Environmental Protection (1) Deforestation, habitat loss, and species extinction are only some of the negative ecological impacts that have been linked to conventional tourism. Sustainable tourism is based on the principles of conservation and pollution prevention.

Carbon emissions are largely attributable to travel, with air travel being the main offender. Reducing emissions of greenhouse gases and encouraging environmentally friendly modes of transportation are two main goals of sustainable tourism.

Thirdly, Cultural Conservation: Unchecked tourism can threaten indigenous ways of life, leading to the watering down and commercialization of their traditions and customs. Responsible tourism is tourism that helps and doesn't hurt the places you visit.

(4) Financial Gains Eco-friendly vacations are good for local economies because they encourage community-based tourism,

support local companies, and generate new jobs. It's a win-win situation because it gets travelers to buy local goods and help out mom-and-pop shops.

5. Empowerment of Local Communities: Empowering local communities to take charge of their own development and conserve cultural heritage is a key benefit of sustainable tourism.

6. Scholarships and Grants Sustainable travel stimulates cultural interchange and provides passengers with chances for learning and personal improvement. It has the potential to encourage people to learn more about other cultures and value their differences.

Multiple Eco-Friendly Vacation Ideas

Everything from mode of transportation to lodging and activities can be considered part of sustainable tourism. Some common methods of eco-friendly vacationing are as follows:

Green modes of transportation include: (1)

- Public Transportation Take buses, trains, and trams to lessen your impact on the environment.

Reduce your carbon footprint and feel more in tune with nature by cycling to your destination.

- Walking: Embrace the simple act of walking to explore cities, landscapes, and wildlife reserves sustainably.

2. Eco-Aware Lodging:

Eco-lodges are lodges that are built and run in a way that causes as little harm to the environment as possible.

Hotels that have taken steps to reduce their environmental impact, such as implementing energy-saving measures, recycling programs, and using local suppliers, are referred to as "green hotels."

Third, "Responsible Wildlife Tourism"

Wildlife Refuges: Seek out places that put animal care and conservation before tourist attractions.
Enjoy wildlife in their native habitats from a distance on guided trips that are both responsible and ethical.

- Homestays: Stay with a local family to get a true feel for the area's customs and lifestyle.
Take part in cultural trips and events organized by local communities.

Reduce the use of single-use plastics 5.

Carry reusable items such as water bottles, shopping bags, and food storage containers with you to cut down on single-use plastics.
To help save the environment, please Refuse Straws and instead use metal or bamboo straws.

6. Aid Regional and Community Economies

- Local Markets: Show your support for independent merchants by frequenting farmers' markets and specialty shops.
Eat Local means eating at eateries that use locally sourced ingredients.

Responsible waste disposal and recycling should be practiced at all times, including when on the road.
Leave no trace; maintain a litter-free environment in parks, forests, and on beaches.

Responsible adventure travel is the eighth topic.

Take part in eco-tours led by reputable companies that put your safety and the health of the environment first.
Follow the guidelines laid out by Leave No Trace while you're out in the wilderness.

Sustainable travel and its revolutionary effects.

Conservation of natural areas, ecosystems, and species variety are all aided by ecotourism since it lessens the toll taken on the environment.

2. Mitigation of Climate Change: Sustainable Travel reduces carbon emissions, which facilitates progress toward climate targets. As the aviation sector works to lessen its impact on the environment, this is of paramount importance.

Thirdly, Cultural Conservation: Sustainable tourism helps locals keep sharing their culture with tourists by encouraging genuine experiences and minimizing negative impacts on the environment.

(4) Financial Gains Eco-friendly vacations are great for local economies, especially small businesses and the people that work there.

5. Empowerment of Local Communities: When locals are included in tourism projects, they gain agency over their own growth and the protection of cultural traditions.

Sustainable travel provides opportunities for learning about other cultures, the environment, and the problems and solutions that are part of the sustainability movement.

7. [Citizen-Supported] Wildlife Tourism Animals are more likely to be respected and preserved if ethical wildlife tourism methods are promoted rather than exploited for entertainment.

8. Reduce the Use of Single-Use Plastics: Sustainable tourism helps the environment by lowering the amount of trash left behind by visitors.

When tourists spend money at local shops, galleries, and marketplaces, that money stays in the communities that hosted them.

Ten. Waste Reduction Careful waste management in the tourism business can have a major effect on lowering pollution levels and protecting natural habitats.

Issues and Things to Think About

While there are many upsides to eco-friendly vacationing, there are also certain things to keep in mind.

Accessibility (No. 1) Not all locations provide eco-friendly vacation options, which can restrict your options.

2. Cost: Some sustainable travel options may be more expensive than typical tourism, thereby restricting access for budget-conscious travelers.

3. Consciousness of the Consumer Travelers must be informed about sustainable travel habits and encouraged to make ecologically conscientious decisions.

Legal Structures: The rules and regulations of governments are crucial in encouraging eco-friendly vacations. It is crucial to lobby for laws that will help the cause.

Transformation of the Industrial Sector: Sustainability and environmentally sensitive practices are something that the whole travel sector, from airlines to hotels to tour operators, needs to adopt.

Conclusion

Sustainable travel is a vital part of the global effort to prevent environmental degradation and climate change. Travelers' actions have far-reaching consequences for the places they visit, the people they meet, and the cultures they experience. Choosing to travel in a way that is less harmful to the environment and future generations is not just an issue of personal preference.

Sustainable travel practices not only enhance our own travel experiences but also help create a healthier, more fair, and more environmentally responsible world for all of us as we chart a course towards a more sustainable future. If we make sustainability a top priority on our travels, we can see the world while doing good for the places we visit and the world as a whole.

aid in the fight to keep the world's cultural and natural wonders intact.

Chapter 5: Sustainable Food Choices

5.1 The Environmental Benefits of Switching to a Plant-Based Diet

Our food habits significantly affect global ecology, population health, and sustainability. Plant-based diets and sustainable eating practices are more important than ever as the globe struggles with climate change and resource depletion. This article of 1,000 words will discuss the importance of plant-based diets, the advantages of sustainable eating to the environment and one's health, and the revolutionary potential of embracing environmentally conscious eating practices.

Grasping the Significance of Vegan Diets

Fruits, vegetables, grains, legumes, nuts, seeds, and other plant-based foods make up the bulk of a plant-based diet, whereas animal products are limited or avoided altogether. There are many reasons why plant-based diets are so crucial:

1. Environmental Conservation: Plant-based diets have a much reduced environmental footprint compared to diets high in animal products. Plant-based food production is more sustainable since it reduces greenhouse gas emissions and habitat degradation.

Greenhouse gas emissions, especially methane from animals, are mostly attributable to animal agriculture. Eating a plant-based diet helps slow climate change by reducing these emissions.

Thirdly, Protecting Biodiversity: Animal agriculture is a major contributor to forest loss, habitat damage, and species extinction. Changing to a plant-based diet alleviates this strain on the environment.

Reducing Water Use and Waste Large quantities of water are needed for cattle and feed crops in order to produce animal-based diets. A plant-based diet uses less water every meal, making it easier on the world's dwindling supply of potable water.

5. Resource Efficiency: Plant-based diets are more resource-efficient as they require fewer resources for feed production, which, in turn, reduces the demand for land, energy, and water.

Health benefits of plant-based diets include a lower chance of developing cardiovascular disease, diabetes, obesity, and several forms of cancer.

Plant-based diets are consistent with ethical concerns about animal welfare since they cause less suffering to animals than conventional diets that are high in animal products.

Sustainable eating has positive effects on both the environment and human health.

The concept of sustainable eating extends beyond the selection of specific foods to incorporate larger ideas that encourage moderate intake. Sustainable eating has many positive effects on our planet, our bodies, and our communities.

One benefit of sustainable eating is a smaller carbon footprint as a result of a diet based on seasonal, locally produced, and minimally processed foods. This is because these foods require less energy to produce and transport.

The second benefit of sustainable eating is the money it brings into the local economy by helping farmers and food producers.

Thirdly, Conservation of Biodiversity is emphasized in sustainable diets by promoting the use of many different types of locally adapted food.

Fourthly, less food waste is the goal of sustainable eating behaviors, as food waste is a major contributor to environmental degradation. It promotes more conscientious shopping, stockpiling, and eating.

5. More Nutritious Options The foundation of sustainable eating is a preference for whole, unprocessed foods and a reduction in the intake of highly processed and sugar-laden goods. Better health and happiness are the results of these decisions.

Sixthly, "Community Development" Sustainable eating habits can improve the vitality of neighborhood farms, increase participation in civic life, and promote the passing on of wisdom and customs.

7. Food Security: Food Security is improved by adopting sustainable eating practices because they encourage the use of a wide variety of locally produced food sources that are less vulnerable to interruptions in global supply networks.

Food Security: Safe food handling and preparation techniques are generally congruent with a sustainable diet, lowering the probability of foodborne disease.

The term "sustainable eating" is broad enough to include a variety of approaches that can be tailored to suit different tastes, schedules, and cultural norms. Important components of environmentally responsible eating include:

One, Foods that are Local and Seasonal: Choose seasonal, regionally-produced foods whenever possible. The environmental impact of long-distance travel is thereby diminished.

Vegan or vegetarian diets: Adopt a plant-based diet or cut back on your intake of meat and other animal products. The ecological effects of plant-based diets are generally lower.

3. Organic and Regenerative Agriculture: Advocate for farming methods that improve soil quality, increase biodiversity, and decrease reliance on synthetic chemicals.

Reduce food waste by correctly planning meals, conserving food, and finding new uses for leftovers; this brings us to tip number four: Minimize Food Waste.

5. Consume Consciously by favoring unprocessed or lightly processed foods over those that have been heavily processed or packaged.

Selecting products with minimal or environmentally friendly packaging is a great way to cut down on trash.

Community Gardens and Allotment Gardens: Grow your own food in communal gardens or on an allotment and develop a closer relationship with the produce you eat.

8. Food Preservation: Master food preservation methods to lengthen the availability of seasonal foods, such as canning, freezing, or drying.

9. Sustained and Regional Seafood: If you eat seafood, pick sustainable options that promote ethical fishing methods and marine ecosystem health.

The Revolutionary Power of Ecologically Sound Diets.

Eating more sustainably has the potential to revolutionize many facets of our lives and the world at large.

1. Environmental Protection: Sustainable Eating is essential in lessening the damage the food business causes to the natural world. It helps slow global warming, protects precious wildlife habitat, and keeps precious materials from running out.

Sustainable eating aids global climate change efforts by decreasing carbon emissions from food production and transportation.

3. Protection of Biodiversity: By encouraging less resource-intensive agriculture, sustainable eating habits lessen the loss of habitats and safeguard biodiversity.

Health and Happiness 4. Better health, lower risk of diet-related diseases, and enhanced well-being result from prioritizing whole, unprocessed foods over highly processed alternatives.

Fifthly, Animal Welfare: Reducing the need for factory farming is consistent with ethical concerns about animal welfare.

6. Support for Local Economy: Sustainable Eating helps local economies by increasing the demand for food that is grown or raised in the area and encouraging the development of family farms.

7. Food security and climate resilience: Encouraging diversified, locally produced food sources promotes food security by lowering dependence on global supply lines.

Issues and Things to Think About

While it's easy to see the positive effects of sustainable eating and plant-based diets, there are still some things to bear in mind.

1) Easily Obtainable and Reasonably Priced Not everyone can afford or has access to sustainable and organic food options.

Respect for ethnic and dietary preferences and the diversity of food traditions around the world are essential components of sustainable eating practices.

3. Consciousness of the Consumer If we want more individuals to embrace environmentally responsible eating habits, we must increase their exposure to sustainable eating practices.

Policies of the Government, Number 4 Government laws and regulations play a significant role in supporting sustainable food production and consumption. It is crucial to lobby for laws that will help the cause.

Conclusion

The global movement to reduce greenhouse gas emissions, safeguard the environment, and improve public health relies heavily on plant-based diets and other sustainable eating practices. Our dedication to the health of the earth and the next generation should inform every aspect of our nutrition, not just our individual tastes.

To leave a healthier, more equitable, and environmentally responsible world for all, adopting plant-based diets and sustainable eating practices is a deliberate choice we can make as we move towards a more sustainable future. It's a reflection of our abilities and a promise to the planet.

to drastically alter our current methods of sustenance and ecological engagement.

5.2 A Sustainable Approach to Nourishment: Reducing Food Waste and Supporting Local Food Sources

Despite its centrality to human existence, food production, distribution, and consumption have far-reaching ecological, financial, and social consequences. Achieving sustainability and food security requires addressing the dual concerns of food waste and supporting local food sources. This essay will discuss the environmental and economic benefits of encouraging local food sources and minimizing food waste, as well as the transformative possibilities of adopting a more sustainable approach to nutrition throughout the course of 1,000 words.

Realizing How Crucial It Is to Cut Down on Food Waste

The effects of food waste are widespread and need to be addressed worldwide. The following essential elements attest to its significance:

1. Conservation of Resources It takes a lot of space, water, and energy to grow food. Food waste is a major contributor to environmental degradation and resource depletion since so much perfectly good food is thrown away.

2. Impact on the Environment: Methane, a strong greenhouse gas, is produced when food waste decomposes in landfills. Throwing away edible food contributes to forest loss, animal habitat degradation, and species extinction.

Thirdly, Food Insecurity: While food is wasted in many parts of the world, millions of people suffer from food insecurity. Eliminating food waste can help feed the hungry and advance social justice by diverting surplus food to those who need it.

Expenses to the Economy The waste of edible food is a major source of financial loss for households and businesses alike. Profitability and savings can both be improved by cutting down on wasted food.

5. Consumer Awareness: Addressing food waste improves awareness about the worth of food, the influence of our decisions, and the need for more sustainable food systems.

The Environmental and Financial Gains from Buying Local

There are several ways in which bolstering regional food production helps ensure the long-term health of local farms and communities.

1) Decreased Impact on the Environment Reduced carbon emissions from long-distance transportation are one benefit of producing and distributing food locally. In other words, this helps fight global warming.

(2) Conservation of Biodiversity: Preservation of Native Crop Varieties and Animal Breeds is generally a top priority in local food systems.

Thirdly, Resource Efficiency: small-scale agriculture uses less water and energy than industrial agriculture on a grand scale.

Boosting local economies, generating new employment possibilities, and fostering the development of new, small-scale farming and food-related enterprises are all benefits of investing in food production close to home (see point #4).

5. Constructing Community By strengthening ties between farmers and diners, local food systems help build stronger neighborhoods. Farmers' markets, CSAs, and restaurants that source their food directly from local farmers all play important roles in fostering these relationships.

Sixthly, Safety and Quality of Food: Produce from nearby farms is typically of greater quality and pesticide-free than that shipped in from further away. Both the food supply and people's health benefit from this.

7. Disaster-Proof Food The local food system is better able to withstand disruptions in the supply chain, environmental calamities, and global crises. They help ensure that people don't go hungry by decreasing reliance on far-off sources.

Efforts to reduce food waste must be made at every stage of the food chain, from cultivation to disposal. Some essential components of minimizing food waste are:

First, How to Reduce Food Waste at Home:

Plan meals in advance to make efficient use of ingredients and cut down on food waste.
- Proper Storage: Store food goods correctly to extend their shelf life.
Use what you have left over in inventive ways to cut down on waste.
To reduce methane emissions and keep organic waste out of landfills, compost food scraps.

Effective Shopping Tip No. 2:

- Buying in Bulk: Get things with a longer shelf life in larger numbers to cut down on packaging and overall waste. - Avoiding Impulse Buys: Shop with a list and don't get carried away.
Learning to read the expiration date on a product and using that information to make purchasing decisions can save you money and time.

The Third Way to Lessen Restaurant and Food Service Waste:

Serve appropriate-sized portions to cut down on food waste.
- Menu Planning: Conceive of dishes that make the best possible use of available ingredients.
Make connections with local food pantries and shelters so you may donate any food that will go to waste.

Food Redistribution and Reclamation System

Participate in a gleaning program that gathers unused farm produce to donate it.

In order to help those in need, donate to your community's food bank.

Initiatives in Policy and Business:

- Legislation: Fight for laws that would limit the amount of food that stores and eateries can throw away.
- Industry Commitments: Prompt food companies to publicly commit to cutting down on food waste and boosting supply chain efficiencies.

A Variety of Options for Promoting Regional Food Systems

Several actions and decisions can help support local food sources by increasing opportunities for local food production, distribution, and consumption. Important factors in bolstering regional food supplies include:

1. Farmers Markets: Visit your neighborhood farmers markets to buy in-season fruits and vegetables straight from the growers.

Second, the concept of "Community-Supported Agriculture" (CSA) You can enhance the bond between consumers and farmers by participating in a community-supported agriculture (CSA) program and receiving a regular allotment of a local farm's products.

Thirdly, Local Food Cooperatives: Join or give to groups that buy primarily from regional growers and makers.

4. Farm-to-Table Restaurants: Dine at restaurants that stress local and seasonal foods in their menus, supporting local food producers.

5. Community Gardens: Participate in community gardening or back programs that encourage the growing of locally sourced food in urban places.

Sixth, Local Food Distribution Networks: Promote the growth and improvement of local food distribution networks to increase the availability of locally produced goods.

Educational Initiatives #7: Advocate for programs that teach people why they should care about their community's food supply and how they can get involved in improving it.

The Revolutionary Power of Fighting Food Waste and Investing in Local Food Systems

There are many ways in which our lives and our society as a whole could be improved by taking steps to reduce food waste and increase the use of local food sources.

1. Effects on the Environment: These actions help to preserve biodiversity, cut down on emissions of greenhouse gases, and save precious resources. They are crucial in the fight against global warming and environmental deterioration.

Reducing food waste benefits both food security and social justice by redistributing surplus food to those in need.

Benefit No. 3: These methods boost local economies, provide new employment opportunities, and promote the development of small-scale farming and food-related enterprises.

Fourthly, "Community Development" Consumers and farmers are brought closer together through the promotion of locally sourced food. Sustainable lifestyles rely heavily on this kind of community development.

Resilience and Food Security 5. Sustainable Eating and Local Food Systems Increase Food Resilience by Reducing Dependence on Global Supply Chains and Fostering Local Self-Sufficiency.

Issues and Things to Think About

First, Educating the Public: More individuals will be encouraged to adopt sustainable behaviors if they are made aware of the problems associated with food waste and the value of investing in local food sources.

Initiatives from government and business also have an important role in

play a crucial part in propelling transformation. It is crucial to lobby for laws that will help the cause.

3. Accessibility and Affordability: It is possible that not everyone has equal access to fresh, local produce and sustainable products.

4. Cultural Preferences: Recognizing the diversity of food traditions across the world, it is important to support local food sources that respect cultural and dietary preferences.

Conclusion

The global effort to prevent environmental degradation, enhance food security, and create more sustainable and resilient food systems relies heavily on reducing food waste and promoting local food sources. The decisions we make about food production, consumption, and distribution have far-reaching effects on the health of our planet, our communities, and the next generation.

Reducing food waste and supporting local food sources is a deliberate action that can have a good effect on the environment, lessen food insecurity, boost local economies, and create more resilient communities as we move forward on the path to a more sustainable future. It's an affirmation of our capacity to effect profound changes in our eating habits and our relationships with others and the natural world.

Chapter 6: Eco-Friendly Consumerism

6.1 Sustainable Futures Nurtured by Ethical Consumption

Shopping is more than just buying things; it's a potent way to show the world what we value and how we want it to be. A more sustainable, egalitarian, and responsible world can be achieved through mindful purchasing and ethical consumerism, which are especially important in today's troubled social and ecological climate. In this 1,000-word article, we will discuss the relevance of conscious buying and ethical consumerism, the environmental and social benefits of these activities, and the transformative potential of adopting a thoughtful attitude to consumption.

Recognizing the Value of Conscientious Consumption

Practicing conscious consumption entails giving our purchases serious consideration before making a purchase. It involves thinking about the effects of our purchases on the planet, society, and ethics. Several crucial elements highlight the significance of mindful purchasing:

One way that conscious consumption helps to preserve scarce materials is by discouraging the purchase of non-essential or disposable items.

2. Impact on the Environment: Everything We Buy Leaves a Footprint on the Environment. By being environmentally conscientious when we shop, we can help sustain a planet in better condition.

3. Social Responsibility: Ethical production methods are promoted by conscientious purchasing, which can result in fair salaries, safe

working conditions, and the preservation of human rights in supply chains.

4. Economic Shift: By promoting goods that share our values, we can increase the demand for environmentally and socially responsible options, thereby influencing the business world for the better.

By supporting the local economy, shoppers show support for their neighborhoods.

Sixth, "Quality, Not Quantity" In the long run, it can save you money by encouraging you to buy high-quality, long-lasting goods rather than cheap, throwaway ones.

The Positive Effects of Ethical Consumption on the Environment and Society

When making purchases, ethical consumers think about how their decisions may affect others and the world around them.

1. Decreased Environmental Impact: Ethical Consumerism drives the purchasing of items with smaller carbon footprints, less waste, and more environmentally responsible sourcing and production methods.

Ethical consumerism helps reduce climate change and greenhouse gas emissions by rewarding businesses that put an emphasis on these issues.

The third benefit of ethical consumption is its impact on environmental protection and preservation through the promotion of eco-friendly goods and processes that minimize waste.

Fair labor practices include the payment of living wages, the provision of health insurance, and the protection of other fundamental worker rights across all stages of production and distribution.

Responsibility to others, number 5. By investing in firms with good morals, we can help underprivileged groups flourish and fund important social initiatives.

Sixthly, "Community Development" In order to strengthen community links and economic resilience, ethical consumers generally choose locally owned enterprises and independent craftspeople.

Seventhly, Quality and Durability: Ethical consumerism encourages the purchase of high-quality, long-lasting items, which helps to minimize waste and promotes environmental sustainability.

Ethical consumerism and mindful purchasing provide a broad spectrum of options that can be adapted to suit each person's values, priorities, and way of life. Some of the most important features of ethical shopping are:

1. Research and Information: Look into the brands you plan to back. Make use of tools such as ethical brand rankings, ethical certifications, and eco-labels.

Environmentally Friendly Products Pick items that are made with long-term environmental health in mind. The use of renewable resources, low-energy requirements, and ethical manufacturing should all be priorities.

3. Ethical Sourcing: Back businesses that put a premium on treating their employees fairly by using ethical sourcing methods and maintaining open communication across the supply chain.

Local Merchandise, Number Four Make purchases from local companies, craftsmen, and farmers a top priority. These options help local economies grow and cut down on pollution caused by transportation.

Buying Fair Trade products ensures that the producers in the supply chain, often located in developing nations, are paid a living wage and are provided with safe and ethical working conditions.

Consider shopping at thrift stores or purchasing used goods to help promote waste reduction and a more sustainable economy.

Minimalism, number seven: Accept minimalism by paring down your belongings and placing value on quality rather than quantity. As a result, less energy is used and thrown away.

Use platforms and apps that inform consumers about the social and environmental effects of the things they're considering purchasing.

The Revolutionary Power of Conscientious Buying and Ethical Consumption

Ethical consumerism and the practice of mindful shopping can have far-reaching effects on people and society.

Environmental Protection (1) To combat climate change and environmental degradation, these methods should be implemented widely because of the significant contributions they make to resource conservation, reduced environmental impact, and a more sustainable approach to consumption.

2. societal Responsibility: Ethical consumption supports fair labor standards, safe working conditions, and human rights in global supply chains, promoting beneficial societal results.

Thirdly, Economic Shift: Conscious Shopping stimulates market demand for more responsible choices by supporting ethical enterprises and sustainable products.

To lessen the negative environmental and social effects of disposable or fast fashion items, conscious consumers can prioritize purchasing high-quality, long-lasting products.

5. Constructing Community Investing in locally-owned companies and craftspeople builds resilient communities and stimulates economic growth.

Issues and Things to Think About

While it's easy to see the positive effects of ethical purchasing and consumption, there are certain obstacles to keep in mind.

First, Educating the Public: In order to get more people to engage in ethical consumption, it is important to educate them about the topic.

Second, there may be inequalities in the availability and affordability of products that are both ethical and sustainable, and this is something that has to be worked on.

Thirdly, Industry Evolution: Changes in the supply chain and in business operations generally are needed to make a more meaningful impact on the environment and society as a whole.

4. Cultural Preferences: Recognizing the Diversity of Consumer Choices and Values, conscious purchasing and ethical consumerism should respect cultural and lifestyle preferences.

Conclusion

The fight against environmental degradation, the promotion of social responsibility, and the creation of a more sustainable, equitable, and responsible world all rely on conscious shopping and ethical consumerism. What we choose to buy and support is a reflection of not only our ideals, but also our agency in shaping the world.

To leave a good imprint on the environment, encourage ethical company practices, support fair labor and human rights, and boost local communities and economies as we move towards a more sustainable future, aware shoppers and ethical consumers may make informed decisions. They are a monument to our power to make dramatic changes in the way we consume, interact with the marketplace, and impact the values and practices of organizations and industries globally.

6.2 "Minimalism and Clutter-Free Living: Streamlining for Environmental Sustainability"

Minimalism and decluttering provide a new lens through which to view the world, freeing us to live more intentionally while leaving less of an impact on the planet. A more sustainable, eco-conscious way of life can be achieved through the practice of adopting a minimalist lifestyle and simplifying our homes. This 1,000-word essay will discuss the relevance of minimalism and decluttering, the advantages to both the environment and the individual, and the possible life-altering effects of making these changes.

Recognizing the Value of Decluttering and Minimalism

The philosophy and way of life known as minimalism is predicated on the idea that less is more. It inspires us to pare down to the essentials in life. To declutter, one must get rid of things that are no longer needed or used. Several crucial considerations highlight the significance of minimalism and decluttering:

1. Conservation of Resources Minimalism encourages thoughtful consumption, which decreases the need for items and hence lowers production costs.

2. Impact on the Environment: Excess items typically end up in landfills, leading to waste and pollution. The negative effects of trash on the environment are diminished because to minimalist practices.

3. Individual Happiness Reduced stress, enhanced mental health, and greater fulfillment in life are all possible outcomes of adopting a minimalist lifestyle. A healthier, happier, and more well-rounded existence is supported by this method.

The Awareness of Consumers 4. Practicing minimalism and decluttering can raise awareness about our habits of consumption and the environmental effects of our decisions.

The Personal and Environmental Benefits of Minimalism

There are numerous personal and planetary benefits to adopting a minimalist lifestyle, such as:

1. Reduced Environmental Footprint: Minimalism leads to fewer purchases and a reduced demand for resources, which, in turn, lowers carbon emissions, waste creation, and habitat damage.

Secondly, Resource Conservation: When you get rid of clutter, you make place for more eco-friendly activities like recycling, composting, and reusing.

The third benefit of minimalism is that it helps keep trash out of landfills by emphasizing quality over quantity.

Reduced stress and enhanced well-being are two direct results of keeping one's living space uncluttered and unmuddled.

5. Increased Focus: By removing the mental clutter associated with unnecessary possessions, minimalism can increase concentration and decision-making.

Monetary Independence 6. Saving money, having more financial security, and more options in life can all result from making fewer purchases and concentrating on what really counts.

Options for Minimalism and Clutter Reduction

Flexibility and personalization are hallmarks of minimalism and other forms of clutter reduction. Some of the most important aspects of these methods are:

First, it's important to prioritize the things and the people in your life based on what you value and what will help you achieve your goals.

2. Sustainable Consumption: Think about the effects of your purchases on the environment and pick eco-friendly, long-lasting products.

Third, practice digital decluttering by cutting down on screen time, sorting out your digital files, and limiting how much media you take in.

Fourth, Rethink Gift-Giving: Ask loved ones to consider giving experiences or simple presents in lieu of material items.

You should get go of something in your home to make room for something new. This regulation aids in keeping the area neat and tidy.

Eco-friendly habits, such as recycling, composting, and reusing products, are in harmony with minimalism and decluttering, so keep them in mind.

Donating or sharing stuff you no longer need encourages a culture of recycling and helps build stronger communities.

There are many facets of our life and our society that may benefit from the revolutionary effects of minimalism and decluttering.

1. Environmental Impact: By decreasing waste, cutting resource consumption, and promoting responsible consumer choices, minimalism and decluttering contribute to a more sustainable and eco-conscious way of life.

2. Conservation of Resources: Reducing environmental deterioration is a top priority, and these methods promote resource-efficient habits including recycling, reusing, and repurposing.

Thirdly, Stress Reduction: Simplifying Our Spaces and Lives with minimalism and decluttering can aid in lowering stress levels and improving one's emotional and psychological health.

4. Independent Financial Status Adopting a minimalist lifestyle can help you save money, gain financial security, and free up time and energy to focus on the things that truly matter in life.

5. Constructing Community Through donation, repurposing, and support of local organizations and charities, minimalism and decluttering can encourage a sense of community and sharing.

Issues and Things to Think About

While the advantages of decluttering and minimalism are clear, there are still obstacles to bear in mind.

Culture of the Consumer (1): Minimalism can collide with the prevalent consumption culture, making it tough to adopt this lifestyle in a society that typically connects goods with success and happiness.

2. Attachment to Possessions: Decluttering might be problematic if you have an emotional attachment to your belongings and find it difficult to let go of them.

Thirdly, Cultural Differences: Minimalism and decluttering should be sensitive to people's varying cultural values and lifestyle preferences.

4. Supportive Infrastructure: Local and national governments should fund and construct recycling centers and donation hubs to encourage citizens to adopt eco-friendly lifestyles.

Conclusion

The principles of minimalism and simplification provide a novel way to reduce stress and improve quality of life. These habits help us simplify our lives, lessen our impact on the planet, and remember what really matters in a culture that often prioritizes materialism and excess.

Embracing minimalism and decluttering is a deliberate action to leave a beneficial impact on the environment, boost personal well-being, and cultivate a culture of awareness and simplicity as we move towards a greener and more sustainable future. It is a tribute to our power to make dramatic changes in the way we consume, engage with our surroundings, and live with intention and purpose.

Chapter 7: Community Involvement

7.1 Joining or starting eco-conscious communities

Eco-friendly neighborhoods shine as examples of how to live sustainably in a world struggling with environmental issues. You can help create a cleaner future by getting involved with an existing community or by founding your own, both of which are examples of eco-conscious societies. In this 1,000-word article, we'll discuss the value of eco-conscious communities, the advantages of participating in them, and the life-changing effects of working together toward sustainability.

Recognizing the Value of Eco-Aware Neighborhoods

Eco-friendly neighborhoods are made up of people who have made a concerted effort to improve their quality of life while also protecting the environment. Several crucial considerations highlight why eco-aware neighborhoods are so crucial:

1. Group Effort Eco-aware groups are more effective when they work together toward a similar goal. Larger initiatives are within their reach, and they have the ability to affect policy and encourage others to adopt sustainable methods.

2. Knowledge Sharing: Eco-aware communities facilitate the exchange of information, insights, and experiences in the field of sustainability. Members can share knowledge and gain access to helpful tools to help them live more sustainably.

Thirdly, emotional upkeep: Maintaining a sustainable lifestyle in our consumer-driven culture can be difficult. In the face of environmental issues, eco-aware societies provide comfort via shared empathy and solidarity.

4. Resource Sharing: Many environmentally aware communities encourage resource sharing since it cuts down on waste, conserves supplies, and fosters a stronger feeling of community.

5 Advocacy and Public Education In order to spread sustainable principles and affect regional and national policy, eco-friendly communities frequently organize lobbying and awareness campaigns.

The Positive Effects of Eco-Aware Communities on the Environment and Individuals

There are numerous environmental and personal benefits that can result from joining or establishing eco-conscious societies.

1. Environmental Impact: These neighborhoods as a whole work to lessen their impact on the environment by cutting back on waste and using as few resources as possible.

2. Resource Conservation: Eco-conscious societies frequently promote recycling, upcycling, composting, and efficient resource use.

3. Renewable Energy: To lessen their dependency on fossil fuels, several municipalities are putting money into solar panels and wind turbines.

4. Local Agriculture: Boosting the local economy and lowering food's carbon footprint by supporting local, sustainable agriculture.

5. Community Development: Eco-aware communities foster a feeling of togetherness and a sense of shared purpose, which improves relationships and well-being.

Sixthly, Schooling: Environmental concerns, sustainable practices, and moral principles are frequently discussed, and members are given numerous opportunity to learn more.

Reduced stress and improved mental health are two additional benefits of belonging to such groups.

A Variety of Options for Participating in or Establishing Eco-Conscious Communities

Numerous opportunities and activities exist that are congruent with sustainable living and shared values, whether you decide to join an existing eco-conscious community or establish your own. Some of the main features of these alternatives are:

First, Pick an Already Existing Community that shares your values and aims through extensive research. Intentional eco-villages, co-housing developments, and local environmental associations are just few of the existing communities.

2. creating a Local Group: If there isn't an existing community in your area, try creating a local group or environmental organization. Bringing together eco-conscious individuals might be as simple as inviting over some friends or inviting over some neighbors.

3. Social Networking Sites: If you'd like to connect with others who share your commitment to sustainability but aren't in close vicinity, you might want to look into eco-friendly online groups.

Sharing equipment, vehicles, and community spaces can lessen the toll on the environment and boost community spirit at the same time.

Engage in community gardening, energy-efficient building renovations, and environmental education programs as examples of collaborative sustainability projects.

6. Advocacy and Awareness: Collaborate on campaigns, events, and activities to raise awareness about environmental challenges and lobby for policy changes.

The Power of Eco-Aware Neighborhoods to Change the World.

There are many ways in which eco-aware neighborhoods might affect positive change:

There has a positive effect on the environment since members of these groups are more likely to adopt eco-friendly practices and to encourage others to do the same.

2. Resource Conservation: Eco-aware communities depend on and benefit from resource-efficient behaviors like recycling, sharing, and urban farming.

Thirdly, Renewable Energy helps communities cut down on their use of fossil fuels while also promoting more eco-friendly power options.

4. Community Building: Eco-aware communities improve interpersonal bonds, instill a sense of belonging, and inspire members to work together toward common goals.

5. Awareness and Education: Sustainability, environmental issues, and ethical living are common topics of discussion between members, and this can lead to a more in-depth awareness of and dedication to sustainable activities.

Influencing Policy and Advocacy Those that care about the environment regularly organize campaigns to promote sustainability at the municipal, regional, and national levels.

Issues and Things to Think About

Joining or establishing eco-conscious communities has clear advantages, but there are also problems and things to think about.

1. Time and Commitment: Joining or establishing an eco-conscious community takes time and dedication because it relies on the efforts of the whole group.

2. Variable Values and Objectives: It's possible for members of an environmentally concerned group to have contrasting priorities and outlooks on how best to achieve their shared goals.

Third, Resource Allocation: Managing the community's funding, resources, and distribution of duties can be difficult, calling for clear and consistent communication and planning.

There may be a need to increase the scope of the influence of eco-conscious communities by working with other groups to bring about greater environmental and social change.

Conclusion

To make a positive mark on the world and encourage collective responsibility as a means to create a more sustainable and fair society, many people are choosing to join or found eco-conscious communities. These communities provide a sense of connection, common ideals, and a shared commitment to greener living in an era marked by environmental difficulties.

Embracing eco-conscious communities is evidence of our ability to make radical changes in our daily lives, relationships with others, and collaborative efforts to create a more just and environmentally friendly world for all.

7.2 Supporting Community Efforts to Improve the Environment by Participating in Clean-Up Events and Tree-Planting Projects

Participating in clean-up and tree-planting efforts is a powerful way to make a tangible influence on your community and the globe, and they are great places to start practicing environmental stewardship. These actions have a positive impact on the local ecosystem and help further ecological restoration efforts. This essay will discuss the environmental and community benefits of participating in local clean-up and tree-planting programs, as well as the transformative potential of such labor, over the course of 1,000 words.

Recognizing the Value of Neighborhood Cleanup and Tree-Planting Efforts

There are many compelling arguments in favor of supporting community-based cleanup and tree-planting efforts:

The attractiveness and cleanliness of neighborhoods are improved almost immediately when these projects are implemented.

2. Environmental Health: Cleanups get rid of trash, garbage, and other debris, which is great for the environment because it helps the air and water quality and the biodiversity.

Third, Restoration: Tree-planting helps with reforestation, which helps restore natural ecosystems and lessens the effects of deforestation.

The absorption of carbon dioxide by trees reduces the severity of climate change and benefits the environment as a whole.

5. Community Engagement: These initiatives encourage participation, unity, and a common dedication to protecting the environment.

There are many environmental and community benefits to getting involved in local clean-up and tree-planting efforts.

1. Cleaner Environment: Efforts to clean up public areas help lower pollution levels and boost environmental quality as a whole.

Enhanced Water and Air Quality These efforts enhance human and environmental health because they reduce pollution and litter in the environment.

Thirdly, Wildlife Habitat: New habitats for wildlife are created when trees are planted, which helps sustain local biodiversity and improves ecosystem health.

4. Climate Resilience: Trees absorb carbon dioxide, which helps to reduce the effects of climate change and makes communities more resistant to the effects of harsh weather.

5. Community Bonding: Taking part in these activities brings people together, helps them meet new people, and gives them something in common to be proud of.

The Sixth: Youth Involvement These programs provide opportunity for youth to become actively involved in environmental efforts, cultivating a feeling of responsibility and stewardship from an early age.

Options for Getting Involved in Community Cleanup and Tree Planting Events

There is a wide range of options for becoming involved in local clean-up and tree-planting programs, depending on one's interests, geography, and available resources. Some of the most important parts of these events are:

Cleaning Efforts: (1) Participate in local park clean-ups, riverbank restoration projects, or litter collecting drives. Initiate your own clean-up effort in a neglected part of your community.

Initiatives to plant trees (2): Join environmental groups, government agencies, and other community groups in planting trees. If you have access to land, you can plant trees as well, so long as you obey the rules and take into account the tree species and growing circumstances in your area.

Third, Educational Outreach: Conduct educational programs alongside clean-up and tree-planting campaigns to increase participation and understanding of environmental issues.

4. Community Collaborations: Establish continuing clean-up and tree-planting efforts that engage a wide variety of stakeholders by partnering with local schools, businesses, and community organizations.

5. Environmental Advocacy: Use clean-up and tree-planting events as opportunities for campaigning for environmental protection and municipal legislation that support these projects.

The Revolutionary Power of Community-Based Environmental Improvement Projects

Taking part in community-wide cleanups and tree-planting events can have far-reaching effects on several levels:

1. Effects on the Environment: Cleaner air and water, more robust ecosystems, and reduced impacts from climate change are all direct results of these efforts.

2. Community Bonding: Participating in environmental work promotes a sense of community, strengthens social relationships, and encourages the development of shared values, which in turn encourages a cooperative and responsible attitude.

3. Youth Empowerment: Youth Participation in Clean-Up and Tree-Planting Activities Encourages Active Stewardship of the Environment and Nurtures a Lifelong Commitment to Sustainability.

4. Educational Value: These exercises provide a practical introduction to environmental challenges and the significance of group effort.

5. Pride in One's Community A cleaner and greener community increases local pride and a shared sense of responsibility for safeguarding the environment.

Participation in clean-up and tree-planting projects can have an impact on municipal policies and priorities, leading to more favorable treatment of environmental issues.

Issues and Things to Think About

Even while helping out with community cleanups and tree planting projects is beneficial, there are some things to keep in mind.

1. Sustainability: These activities must be sustainable and integrated into community practices to have a lasting impact.

2. Resource Allocation: It may be difficult to manage funding, resources, and the distribution of duties, therefore good communication and planning are essential.

3. Participation in the Community: Organizers should anticipate dips in enthusiasm and motivation while trying to increase and sustain community participation.

4. Local Conditions: When planning a tree-planting initiative, participants should think about the local environment, climate, and tree species that would thrive.

Conclusion

One way to leave a lasting mark on the world, strengthen local communities, and help create a cleaner, greener environment is to take part in local clean-up and tree-planting efforts. These events provide a way to put your concerns about the environment into action, connect with people who feel the same way, and have a real impact on your neighborhood.

Local clean-up and tree-planting initiatives demonstrate our capacity to make profound changes in environmental care, community building, and the creation of a healthier and more beautiful world for future generations as we travel the path towards a more sustainable future.

Chapter 8: Advocacy and Political Engagement

8.1 Advocacy's Role in Resolving Environmental Crises Through Lobbying for Climate Policies

One of the most urgent problems we face today is climate change, which necessitates coordinated efforts on a global, national, and even community scale. Lobbying for climate policies is an important means of influencing government choices and impacting the future of environmental protection. We will discuss the need of lobbying for climate legislation, the positive effects of advocacy on the environment and society, and the revolutionary potential of using political power to address climate challenges in this 1,000-word essay.

Recognizing the Role of Lobbying in Influencing Climate Policy

There are several compelling arguments in favor of climate change advocacy.

1. Political Influence: Lobbying allows individuals, organizations, and businesses to persuade policymakers to make climate change mitigation a top priority.

New Policy: Successful lobbying can lead to emission regulations, protections for natural resources, and the promotion of sustainable activities; all of which are crucial to addressing climate change.

Thirdly, Public Awareness is increased through lobbying, resulting in a surge of pressure on politicians to address climate change.

4. International Partnership International accords can be influenced by lobbying, and it can inspire countries to work together to address global climate concerns.

Positive Effects of Climate Policy Lobbying on the Environment and Society

There is a wide range of environmental and socioeconomic benefits to climate policy advocacy, including:

Reduced emissions of greenhouse gases are one way in which the effects of climate change can be mitigated or at least slowed.

2) Conservation of Resources: Efforts to advocate for change can lead to better resource management, which in turn reduces deforestation, overfishing, and habitat loss.

3. Renewable sources of power: Reducing reliance on fossil fuels and facilitating the transition to sustainable energy systems, lobbying can help promote the use of renewable energy sources.

Protecting natural ecosystems and wildlife habitats is one way in which climate policy can help with biodiversity preservation.

5. Economic Opportunities: Climate measures frequently boost green industries, producing jobs and supporting economic growth through innovation and sustainability.

Global action on climate change requires international cooperation, which can be fostered through effective lobbying.

A Variety of Lobbying Methods for Climate Policy

Lobbying for climate legislation can be done in a variety of ways, each according to the tastes, resources, and ends of the advocate. Some of the most important parts of these plans are:

1. Interacting with Legislators: Meet with, write, email, or call your representatives and senators to voice your concerns and urge them to support measures to address climate change.

2. Grassroots Campaigns: Participate in or create campaigns that aim to educate the public, rally citizens, and show popular support for climate action.

Third, collaborate with environmental groups that already have the know-how and means to mount successful advocacy campaigns.

Increase public understanding of climate change by distributing educational materials and hosting activities in the neighborhood.

To maximize the effectiveness of advocacy efforts, it is important to forge partnerships with other like-minded groups and individuals.

Investor activism #6: advocate for fossil fuel divestment and greener investing strategies.

The Game-Changing Power of Climate Policy Lobbying

There are many facets of our lives and our society that may be revolutionized by climate change advocacy.

1. Environmental Impact: Advocacy activities support the establishment and implementation of climate policies that directly effect environmental circumstances, such as decreasing emissions, conserving natural resources, and fostering renewable energy.

Second, Resource Conservation: Climate policies can promote healthier ecosystems by encouraging responsible resource management, habitat protection, and biodiversity conservation.

Economic Opportunities 3. Advocacy can encourage the emergence of green companies, which can provide jobs and promote economic development through sustainable practices.

Global Cooperation: Effective Lobbying Efforts Can Lead To International Cooperation And Commitments To Address Climate Change On A Global Scale, Promoting Collective Action.

5. Public Awareness: Climate Advocacy educates the general public on the issue of climate change and inspires people to take action in response to environmental threats.

Issues and Things to Think About

While it's clear that advocating for climate regulations will help in the long run, there are still obstacles to bear in mind.

Lobbying for climate legislation is typically met with political opposition, which can test one's ability to persevere despite obstacles.

To be successful in lobbying, advocacy campaigns need resources including money, knowledge, and organizational backing.

It can be difficult to make head or tails of climate policy, let alone advocate for specific solutions.

Differing viewpoints: Advocacy activities frequently involve varied stakeholders with varying viewpoints, necessitating adept communication and teamwork.

Conclusion

The decision to lobby for climate policy is an active step toward influencing the future of environmental protection and leaving a constructive legacy for future generations. Advocacy is a potent tool for swaying political decision-makers and bringing about transformative change in a society that is more aware of the necessity of tackling climate change.

Advocating for climate legislation shows that we have the wherewithal to effectively address environmental concerns as we chart a course towards a more sustainable future. It's a way to influence policy and safeguard our planet so future generations may inherit a place that's better for their health and the planet's.

8.2 Climate activism and grassroots movements: growing change from the bottom up

One of the greatest issues of our time is climate change, which calls for concerted, immediate action from all sectors of society. Individuals, communities, and local groups have emerged as potent forces in the global struggle against environmental degradation through participation in grassroots movements and climate activism. In this 1,000 word essay, I want to discuss the significance of climate activism, the advantages it has for society and the environment, and the possibility for change it may bring about in the face of global challenges.

The Value of Climate Activism and Grassroots Movements

There are many reasons why grassroots organizing and climate advocacy are so important:

Participation in the Community: (1) They encourage people to take responsibility for the climate by giving them the tools to act locally.

2. Change Advocacy Movements at the grassroots level put pressure on governments and businesses to adopt climate-friendly policies and practices.

Awareness and Education: Climate Activism spreads information about the seriousness of global warming and encourages people to adopt more eco-friendly lifestyles.

4. International Cooperation Many grassroots efforts unite people from many countries with the common goal of addressing climate concerns.

The environmental and social benefits of taking part in climate activism and grassroots movements are numerous.

Reducing emissions of greenhouse gases is one way to lessen the severity of climate change and its effects.

2. Conservation of Resources: These reforms encourage prudent resource administration and thereby cut down on activities like forest clearance, overfishing, and the degradation of natural habitats.

3. Biodiversity Preservation: Climate activism helps safeguard natural ecosystems and wildlife habitats, supporting biodiversity protection.

Economic Prospects 4. Sustainable innovation and job creation are two key outcomes of grassroots efforts that boost environmentally friendly industries.

5. Community Building: Engaging in climate activism helps strengthen local communities by promoting friendships and common goals.

6. Youth Engagement: Grassroots movements inspire today's youth to take an active role in protecting the planet for future generations.

A Variety of Strategies for Grassroots Climate Activism

It is possible to tailor one's participation in grassroots movements and climate activism to one's own interests, geographical location, and available resources. Elements crucial to these measures are:

First, Local Campaigns: Get involved with or start your own local clean energy, waste reduction, or climate education campaigns.

2) Community Organizing: Plan local get-togethers, classes, and other activities to educate people on climate change and inspire them to take action.

3. Advocacy and Outreach: Collaborate with like-minded organizations to increase the impact of advocacy activities and lobby local and national policymakers for climate-friendly policies.

Create and take part in educational activities that inform people about climate change and the importance of eco-friendly lifestyle choices.

5. Art and Culture: Involve the public in climate activism by utilizing artwork, music, and cultural events.

6. Protests and Demonstrations: Plan and take part in rallies and other public displays to bring attention to the climate crisis and encourage policymakers and business leaders to take action.

The Power of Local Efforts and Climate Activism to Change the World

Change can come about in several forms as a result of grassroots movements and climate activism:

1. Effects on the Environment: By lowering emissions, preserving natural resources, and promoting the use of renewable energy,

these programs have a direct impact on the state of the environment.

Responsible resource management, habitat preservation, and biodiversity protection are all fostered by grassroots initiatives that aim to protect natural areas.

(3) Economic Opportunities: Activism can support the establishment of green companies, contributing to job creation and fostering economic development through sustainability.

4. Global Solidarity: Effective grassroots activism generates international cooperation and collaboration in addressing climate change on a global scale.

5. Public Awareness: Climate Activism educates the public about climate change and inspires people to take action in response to environmental threats.

Issues and Things to Think About

Although the advantages of climate activism and grass-roots movements are clear, there are still obstacles to bear in mind.

Political Opposition, No. 1: Political opposition is common for climate activists, so keeping up the fight requires perseverance and a positive attitude even when things don't go as planned.

Allocating resources such as money, knowledge, and organizational backing is essential for successful advocacy campaigns.

Thirdly, Variety of Stakeholders Many people with different perspectives need to work together effectively in grassroots initiatives.

4] Cultural Sensitivity: Recognizing the diversity of approaches to climate activism, advocacy campaigns should respect cultural values and beliefs.

Conclusion

Activism at the grassroots level and efforts to mitigate climate change are deliberate strategies for making a lasting difference in the world and shaping the future of environmental protection. The importance of grassroots activities in the face of the growing global recognition of the necessity of tackling climate change

movers and shakers, sparking positive change by harnessing the potential of individuals.

Movements from the bottom up and climate activism show that we can make a difference in solving environmental problems as we head towards a more sustainable future. They're a way to provide people the tools they need to build a world that's better for everyone now and in the future in terms of health, sustainability, and resilience.

Chapter 9: Green Technology and Innovation

9.1 Sustainability in the Future, Made Possible by Recent Developments in Green Technology

Advances in green technology have emerged as a ray of hope for a more sustainable future in the face of urgent environmental issues. With the pressing need to combat climate change and environmental damage, new technologies are revolutionizing our daily lives. This 1,000-word paper will discuss the relevance of green technological advancements, the positive effects these developments have on the environment and society, and the revolutionary potential they have for creating a more environmentally friendly and sustainable society.

Recognizing the Value of Innovations in Green Technology

Improvements in environmentally friendly technologies are crucial for many factors.

These developments provide strategies for lowering greenhouse gas emissions, cutting down on pollution, and conserving natural materials.

2. Sustainability: Green technology encourages sustainable behaviors, which match human actions with the ability of Earth's ecosystems to replenish resources and absorb waste.

3. Economic Growth: The green technology industry generates economic opportunities, which in turn promotes innovation, the production of new jobs, and overall economic expansion.

Energy efficiency is another area where green technology has made strides, cutting down on carbon emissions and other negative effects on the environment.

5. Resilience: Communities and infrastructure can be made more resistant to the effects of climate change and other environmental challenges with the help of green technology.

Benefits to Society and the Environment from Recent Green Technology Developments

The first benefit of using green technology is that it helps lessen the impact of climate change by decreasing emissions of greenhouse gases.

2. Conservation of Resources: Deforestation, overfishing, and habitat loss can all be minimized because to these technological advancements, which encourage careful resource management.

Thirdly, the quality of the air and water is enhanced by the use of green technology, which lessens the impact on the environment caused by a number of different operations.

Sustainable practices encourage biodiversity conservation, which safeguards natural ecosystems and the homes of species.

Economic Opportunities 5 The green technology sector encourages economic growth, employment creation, and innovation in sustainable industries.

Advances in green technologies inspire cooperation across borders to solve environmental problems around the world.

The Range of Innovations in Green Technology

Green technology developments span a broad spectrum, allowing them to be honed to meet individual environmental and societal needs. Some of the most important features of these advances are:

1. Renewable Energy: The increased use of renewable energy sources like solar, wind, and hydropower can lessen our reliance on finite resources and cut down on harmful emissions.

2. Energy Efficiency: Innovations in energy-efficient technology for homes, transportation, and industry cut energy consumption and greenhouse gas emissions.

Thirdly, "Green Building" refers to the use of eco-friendly and cost-effective methods of construction.

Sustainable Agriculture: Recent developments in organic farming and precision agriculture, for example, lessen the ecological toll of feeding the world.

5. Waste Reduction: Technologies for waste reduction, recycling, and waste-to-energy conversion lessen the negative effects of trash pick-up on the environment.

Water management is improving water conservation and pollution avoidance through new technology for treating wastewater and using water more efficiently.

To minimize emissions and reliance on fossil fuels, the transportation sector is undergoing a radical transformation, thanks to innovations like electric automobiles, efficient public transportation, and alternative fuels.

Circular Economy, number 8. Circular economies aim to reduce waste and pollution by emphasizing conservation of resources and recycling/reuse of materials.

New eco-friendly technologies may drastically alter many facets of our life and culture.

1. Effects on the Environment: These developments have direct effects on environmental conditions, improving them by decreasing emissions, preserving natural resources, and increasing the use of renewable energy.

2. Resource Conservation: Green technology fosters responsible resource management, habitat protection, and biodiversity conservation, supporting healthier ecosystems.

Thirdly, Economic Opportunities arise from the expansion, employment creation, and innovation in sustainable industries that are propelled by the development of green technology.

In order to effectively address climate change and other global environmental concerns, nations must work together on effective green technology solutions.

5. Public Awareness: As more people become exposed to the benefits of green technology, they become more aware of the importance of taking action to combat climate change and environmental damage.

Issues and Things to Think About

While there is little doubt about the positive effects of green technology advancements, there are still some obstacles to bear in mind.

Investment and Costs 1. Some people and businesses are put off by the initial investment and potential greater costs associated with adopting green technology.

Technical Restrictions Some environmentally friendly technologies are still in the experimental stages, thus they may not be as reliable or affordable as their more established counterparts.

Thirdly, Behavior Change: It can be difficult to obtain widespread acceptance of green technology since it typically necessitates a change in behavior and practices.

Legal Structure(4): Adoption of green technology can be aided and encouraged by a well-thought-out regulatory framework. Regulations that are either inconsistent or inadequate can slow development.

Conclusion

Investing in green technology is a deliberate act with far-reaching consequences for the future of environmental conservation. These developments provide a way forward toward a more environmentally friendly, sustainable, and robust society in a period marked by environmental challenges.

Our ability to make revolutionary shifts in how we live, work, and connect with the planet is reflected in the rapid development of green technology as we chart a course toward a more sustainable future. They help solve global problems, boost the economy, and make the world a better, more sustainable place for future generations.

9.2 Greener Futures through Sustainable Building and Architecture

A more eco-friendly and resilient future can be achieved through the implementation of sustainable building and architecture methods, which have gained prominence in a time of rising environmental awareness and the necessity of addressing climate change. These methods are revolutionizing the way we design, create, and live our built world. In this 1,000 word essay, I want to discuss the importance of eco-friendly architecture and construction methods, as well as the positive effects these developments have on the environment and society.

Insight into the Value of Eco-Friendly Construction Methods

There are many compelling arguments in favor of environmentally responsible design and construction:

1. Responsibility to the Environment Reducing carbon emissions, conserving resources, and encouraging appropriate land use are top priorities in these methods.

Second, Climate Mitigation: Green buildings help reduce global warming by using less power and producing fewer greenhouse gases.

Thirdly, Resource Efficiency refers to the efforts made by eco-friendly architects and builders to make the most of their materials while also reducing waste and bolstering the ideas behind the circular economy.

4. Health and Happiness of Humans: Better indoor air quality and more windows allow more natural light into sustainable buildings, making them more pleasant places to live and work.

5. Resilience: Features incorporated into sustainable architecture help make structures more resistant to the effects of natural disasters and other climate-related stresses.

Sustainable building and design practices have positive effects on both the environment and society.

Sustainable building and architecture methods have many positive effects on the environment and on society.

1. Emission Reduction: Greener buildings consume less energy and produce fewer greenhouse gases, aiding the cause against global warming.

2. Conservation of Resources: Efficient use of water, energy, and materials are just a few examples of how these methods benefit society.

3. The Quality of Our Air and Water Sustainable buildings generally contain technologies and designs that improve indoor air quality and limit pollutants, improving human health.

Fourthly, Biodiversity Conservation is promoted by sustainable design since it safeguards ecosystems and wildlife habitats.

5. Economic Opportunities: Sustainable construction is a driver of economic growth, innovation, and opportunity.

Creating public spaces, improving the visual appeal of the neighborhood, and encouraging community involvement are all ways in which sustainable buildings contribute to community development (see also point #6).

Innovations in Sustainable Building and Architecture

To reduce negative effects on the environment and encourage responsible development, sustainable building and architecture techniques incorporate many different ideas and principles. Some of the most important aspects of these methods are:

Sustainable structures use less energy because they have more efficient lighting, heating, cooling, and insulation.

The use of solar panels, wind turbines, and other renewable energy sources in construction can significantly cut down on the amount of fossil fuels needed to power a structure.

Materials that are recyclable, reusable, or sourced in a sustainable manner are given top priority in green building.

4. Passive Design: Passive design concepts maximize passive heating and cooling by maximizing natural ventilation, daylighting, and thermal performance.

5. Water Efficiency: Water-saving solutions, such as low-flow fixtures, rainwater harvesting, and wastewater treatment, are commonplace in sustainable buildings.

Sixth, "Green Walls and Roofs" These additions boost efficiency, better the environment, and create more room for vegetation.

Reusing structures for new purposes reduces waste and protects historic structures through adaptive reuse.

The goal of sustainable site design is to lessen the negative effects of building on the environment while enhancing natural landscapes.

The Game-Changing Power of Eco-Friendly Construction and Design.

The adoption of more eco-friendly building methods has the potential to revolutionize many facets of modern life:

1. Effects on the Environment: These developments have direct effects on environmental conditions, improving them by decreasing emissions, preserving natural resources, and increasing the use of renewable energy.

Responsible resource management, habitat protection, and biodiversity conservation are all fostered by sustainable building and design practices, which leads to better ecosystems.

Opportunities for economic growth, employment creation, and technological advancement in environmentally friendly businesses are all facilitated by sustainable construction practices.

Creating public spaces, improving the visual appeal of the neighborhood, and encouraging community involvement are all ways in which sustainable buildings contribute to community development.

5. Human Health and Well-Being: Sustainable structures create more pleasant places to live and work, which is beneficial to people's emotional and physical health.

In order to better withstand the effects of climate change and other environmental issues, sustainable architecture incorporates design elements that increase resilience.

Issues and Things to Think About

The advantages of eco-friendly construction and design are clear, but there are still some things to think about.

Investment and Costs 1. Some people and businesses may be put off by the higher initial investment required for sustainable building projects. While there may be some up-front costs, the money saved on energy and maintenance over time usually makes up for it.

Technical Restrictions It's possible that sustainable building materials and technology won't be as readily available or cheap as more conventional options just yet.

Thirdly, the Regulatory Structure: Effective rules and incentives are needed to promote and encourage the implementation of sustainable construction and architecture techniques. Regulations that are either inconsistent or inadequate can slow development.

Training and Education #4: The widespread adoption of sustainable building concepts depends on an educated workforce in sustainable construction methods.

Conclusion

Consciously opting for environmentally friendly building methods is one way to determine the future of environmental protection and leave a lasting legacy. These methods can help us create a constructed environment that is more sustainable, robust, and visually beautiful even as the world as a whole faces serious environmental issues.

Sustainable building and architecture practices are an example of how far we've come in our ability to radically alter the ways in which we plan, build, and live in our urban environments. They help solve global problems, boost the economy, and make the world a better, more sustainable place for future generations.

Chapter 10: The Path Forward

10.1 Setting personal and collective sustainability goals

Setting personal and community sustainability objectives has become a critical strategy of crafting a more eco-friendly and sustainable future in an era characterized by environmental concerns and the urgency of addressing climate change. Whether we're working on them separately or as a group, these objectives are crucial for keeping us on track, encouraging responsibility, and propelling progress. In this 1,000-word essay, I'll discuss why it's important to set individual and societal sustainability objectives, what those goals can accomplish for the environment and society, and how they can help usher in a more sustainable future.

Realizing the Value of Sustainable Objectives

For several reasons, it is crucial for individuals and communities to establish sustainability targets:

1. Responsibility to the Environment These objectives place an emphasis on individuals and communities working together to lessen their environmental footprint, cut carbon emissions, save resources, and advance sustainable methods of operation.

Climate change mitigation is aided by sustainability goals since they cut back on fossil fuel use, GHG emissions, and trash accumulation.

Goal-setting promotes resource efficiency by reducing waste and bolstering circular economy concepts.

4 Human Health and Well-Being: Sustainable Practices directed by Goals produce healthier and more comfortable living and working environments with enhanced indoor air quality and natural light.

5. Resilience: Sustainable development objectives strengthen the ability of people and systems to withstand the negative effects of climate change and other environmental stresses.

Setting sustainability goals has positive effects on the environment and society.

There are many positive effects on the environment and on society as a whole when individuals and communities adopt sustainability goals.

1. Emission Reduction: Sustainability Goals help in the fight against climate change by decreasing energy use and emissions of greenhouse gases.

Second, Conservation aims to encourage the wise use of resources including water, energy, and materials.

Air and Water Quality: Sustainability Goals frequently include technology and practices that enhance indoor air quality, decrease pollution, and lessen the negative impact of various operations on the environment.

4. Conservation of Biodiversity Communities and individuals motivated by sustainability goals are more likely to make decisions that safeguard ecosystems and wildlife habitats, thereby helping to preserve biodiversity.

5. Economic Opportunities: The pursuit of sustainability goals offers economic opportunities, which in turn fosters innovation, the development of new jobs, and overall economic growth.

6. Community Development: Goal Setting encourages local engagement in sustainability activities and creates shared ideals.

Goals for Sustainable Living: Individual and Group Efforts

The process of establishing sustainability objectives involves a wide range of individual and group activities that can be adapted to meet a variety of environmental constraints and human requirements. Elements crucial to these measures are:

1. Modifications to One's Own Way of Life Personal sustainability targets might include behavior modifications including cutting back on energy use, recycling more, and switching to a more eco-friendly mode of transportation.

Second, through Community Engagement, groups can work together to improve their local environment by establishing sustainability goals, such as hosting clean-up days or planting a community garden.

Third, Company and Group Dedications: Companies and organizations can work toward sustainability by establishing targets like cutting down on emissions, spreading eco-friendly policies, and adopting renewable energy sources.

Initiatives in Education (4th) Educational institutions can set sustainability goals by including environmental education into their curricula and promoting sustainable habits among students and staff.

Advocacy and Outreach: Advocacy groups can establish sustainability goals to educate the public on environmental concerns, lobby for legislative changes, and rally support for eco-friendly actions.

6. Technology and Innovation: Innovators and business owners can drive the creation of environmentally friendly products and services by establishing sustainability targets.

Setting sustainability targets can have a profound effect.

There are many facets of our life and our society that could undergo radical change if we set sustainability goals.

1. Effects on the Environment: These objectives have a direct impact on environmental quality by lowering emissions, protecting scarce resources, and encouraging long-term stewardship.

Conservation of Resources and Biodiversity are all boosted by defining goals; hence, ecosystem health improves.

Thirdly, Economic Opportunities arise as progress is made toward sustainability goals, which in turn drives economic growth, job creation, and innovation in sustainable industries.

Creating shared values, improving the visual appeal of neighborhoods, and encouraging local participation are all ways in which the pursuit of sustainability goals contributes to community development.

Five, Human Health and Well-Being: Sustainable practices led by sustainability goals provide healthier living and working environments, which is beneficial to people's emotional and physical health.

6. Resilience: Communities and individuals are better able to withstand the effects of climate change and other environmental difficulties when they adopt sustainability goals.

Issues and Things to Think About

While it's clear that setting sustainability targets will pay off in the long run, there are still certain obstacles to bear in mind.

Altering One's Way of Behaving Changes in attitude and routine are typically necessary for the successful introduction of sustainability objectives.

Allocation of Resources Funding and organizational backing may be required to make investments and allocate resources toward achieving sustainability objectives.

Thirdly, the Regulatory Structure: The adoption of sustainability goals can be supported and encouraged with the use of effective rules and incentives. Regulations that are either inconsistent or inadequate can slow development.

Respect for cultural values and ideas is essential to achieving sustainability goals and helps to acknowledge the variety of perspectives on the topic.

Conclusion

In order to leave a positive imprint on the world, encourage responsible behavior, and determine the future of environmental preservation, setting personal and communal sustainability objectives is a smart move. These objectives can help us move forward together toward a more sustainable, resilient, and peaceful society in an era defined by environmental difficulties.

Setting sustainability goals as we travel the road to a more sustainable future is evidence of our power to effect lasting change in our daily lives, workplaces, and interactions with the earth. They help solve global problems, boost the economy, and make the world a better, more sustainable place for future generations.

8.2 A Path Toward Environmental Stewardship: Encouragement and Call to Action

In an age marked by rising environmental difficulties, the power of encouragement and a compelling call to action are important in moving individuals, communities, and civilizations towards environmental stewardship. Climate change, biodiversity loss, and pollution are all challenges that are gaining attention, but it is encouragement and a call to action that will lead to real solutions. The environmental and societal benefits of encouragement and a compelling call to action, as well as the revolutionary potential they hold in tackling critical environmental concerns, will be discussed in this 1,000-word article.

Recognizing the Value of Inspiration and a Call to Action

Reasons why encouragement and a call to action are so important include:

Motivation: (1) Encouraging people and groups to assume environmental responsibility and work toward sustainable practices.

The second step in solving environmental problems is raising public knowledge of those problems and the magnitude of their effects.

Mobilization (3): These inspiring resources get people and groups working together to achieve environmental goals.

4. Inclusivity: People of various ages, ethnicities, and socioeconomic backgrounds can participate in environmental stewardship provided they are given the support and inspiration to do so.

The Positive Effects of Motivation and a Call to Action on the Environment and Society

The environmental and societal benefits of encouragement and a compelling call to action are numerous.

Reduced Emissions: 1. Encouragement and a call to action can motivate people to take real, actionable efforts toward lowering their carbon footprint and halting climate change.

2. Conservation of Resources: Motivated by these factors, communities might implement water, energy, and material conservation programs.

3. Biodiversity Conservation: Communities can be inspired to safeguard natural ecosystems and wildlife habitats via the use of encouragement and a compelling call to action.

4. Youth Engagement: Motivated by encouragement and a call to action, the younger generation can become active stewards of the environment, developing a lifelong commitment to sustainability.

5. Community Building: Environmental Initiatives based on encouragement and a call to action strengthen community ties, promote social connections, and establish common ideals among individuals and groups.

6. International Cooperation: Motivated by these forces, people all across the world can work together to solve the planet's most pressing environmental problems.

Changes in the Environment and Society as a Result of Inspiration and a Call to Action

A call to action and words of encouragement can have a profound impact on many facets of our lives and the world at large.

1. Effects on the Environment: These incentives have a direct impact on environmental quality because they cause people to produce less waste, use fewer natural resources, and act more responsibly.

2. Resource Conservation: Encouragement and a compelling call to action support responsible resource management, habitat protection, and biodiversity conservation, building healthier ecosystems.

Economic Possibilities 3. These incentives can help push people and places to invest more heavily in environmentally friendly companies, which in turn will create jobs and promote long-term economic growth.

For international collaboration and promises to address climate change and other global environmental concerns, strong encouragement and a compelling call to action are necessary.

5. Youth Empowerment: Motivational Efforts encourage the next generation to get involved in environmental projects, creating a generation that is permanently dedicated to preserving the planet.

Promote local involvement in sustainability efforts through the development of community relationships, the strengthening of shared values, and the promotion of social connections through encouragement and a compelling call to action.

Issues and Things to Think About

The advantages of inspiration and a call to action are clear, but there are obstacles and things to think about.

Social Disinterest 1. Overcoming environmental apathy and opposition to change can be tough. People who aren't invested in environmental issues need to be motivated in productive ways.

The Art of Communicating The call to action and words of encouragement must be simple, understandable, and accessible to a wide range of listeners.

Allocating resources to initiatives that are driven by these factors is a third consideration.

Recognizing the variety of people's approaches to environmental stewardship, it is important that calls to action and other forms of encouragement accommodate cultural values and beliefs.

Conclusion

In a time of environmental difficulties, rising consciousness, and the pressing need to confront climate change, encouragement and a strong call to action are essential. They are an effective way to inspire people to become better environmental stewards, spread awareness about the importance of responsible behavior, and work toward a more sustainable future.

We can prove our ability to inspire change, strengthen community ties, and leave the planet a better place for future generations by offering words of encouragement and a compelling call to action as we make our way towards a greener and more sustainable future.

Printed in the USA
CPSIA information can be obtained
at www.ICGtesting.com
LVHW021807290524
781185LV00010B/82

9 789358 683776